Cost Estimating Manual for Pipelines and Marine Structures

Cost Estimating Manual for Pipelines and Marine Structures

Editor

Rajender Nagpal

Cost Estimating Manual for Pipelines and Marine Structures

Edited by **Rajender Nagpal**

Printed in 2017

ISBN: 978-1-68117-351-1

Library of Congress Control Number: 2015939264

© 2016 by

SCITUS Academics LLC,
616, Corporate Way, Suite 2, 4766,
Valley Cottage, NY 10989

www.scitusacademics.com

Notice

Reasonable efforts have been made to publish reliable data and views articulated in the chapters are those of the individual contributors, and not necessarily those of the editors or publishers. Editors or publishers are not responsible for the accuracy of the information in the published chapters or consequences of their use. The publisher believes no responsibility for any damage or grievance to the persons or property arising out of the use of any materials, instructions, methods or thoughts in the book. The editors and the publisher have attempted to trace the copyright holders of all material reproduced in this publication and apologize to copyright holders if permission has not been obtained. If any copyright holder has not been acknowledged, please write to us so we may rectify.

Contents

Preface

This manual has been compiled to provide time frames, labor crews and equipment spreads to assist the estimator in capsulizing an estimate for the installation of cross country pipelines, marshland pipelines, nearshore and surf zone pipelines, submerged pipelines, wharfs, jetties, dock facilities, single-point morring terminals, offshore drilling and production platforms and equipment and appurtenances installed thereon. The time frames and labor and equipment spreads which appear throughout this manual are the result of many time and method studies conducted under varied conditions and at locations throughout the world; these time frames and labor and equipment spreads reflect a complete, unbiased view of all operations involved. When one is engaged in compiling an estimate from any information furnished by others, as is the case with this manual, he should view it in an objective light, giving due consideration to the nature of the project at hand and evaluating all items that may affect the productivity of labor and all other elements involved.

Editor

Chapter 1

Evaluating EML Modeling Tools for Insurance Purposes: A Case Study

Mikael Gustavsson, Mohammad Shahriari, and
Mats Lindgren

Department of Product and Production, Chalmers University of
Technology, 41296 Gothenburg, Sweden

ABSTRACT

As with any situation that involves economical risk refineries may share their risk with insurers. The decision process generally includes modelling to determine to which extent the process area can be damaged. On the extreme end of modelling the so-called Estimated Maximum Loss (EML) scenarios are found. These scenarios predict the maximum loss a particular installation can sustain. Unfortunately no standard model for this exists. Thus the insurers reach different results due to applying different models and different assumptions. Therefore, a study has been conducted on a case in a Swedish refinery where several scenarios previously had been modelled by two different insurance brokers using two different softwares, ExTool and SLAM. This study reviews the concept of EML and analyses the used models to see which parameters are most uncertain. Also a third model, EFFECTS, was employed in an attempt to reach a conclusion with higher reliability.

INTRODUCTION

The petroleum refineries share their inherent safety problems with many other chemical processing industries. The raw material as well as almost all of the products are highly flammable, can give rise to vapour cloud explosions (VCEs), and are toxic above a certain thresholdvalue. Much of the processing, as well as storage, is done under higher than ambient pressure, not uncommonly above 10 bars. This ensures that loss of confinement will lead to rapid discharge rates.

In addition to the threat to the workforce an economical risk is associated with processing flammable compounds. not only due to the direct impact of a fire or explosion but also the cost of business interruption (BI) in case of a shutdown. In many cases the BI after an accident is much more expensive than the actual repair costs due to a fire or explosion.

Today the highest reported property damage (PD) is from a VCE that occurred in Pasadena Texas 1989. It is estimated that the costs to rebuild the plant were around 869 million USD (based on a 2002 USD) [1]. The BI cost in this case was a mere 700 million USD. Notice here that the BI and PD costs are roughly the same whereas the average, calculated from 119 accidents, is that BI exceeds PD with a factor 2.7 [2].

A decision has to be made by the operator. How much of my financial risk do I want an outside party to carry, and how much money do I consider is a fair price for that service? The decision process generally includes modelling of various scenarios to determine to which extent the process area can be damaged. On the extreme end of modelling the so-called Estimated Maximum Loss (EML) scenarios are found. These models try to predict the maximum loss a particular installation can sustain due to an accident. Within the refinery industry these scenarios usually consist of a number of different vapour cloud explosions. For obvious reasons such scenarios are riddled with uncertainties. Some scenarios are frowned upon by some and deemed plausible by others.

Unfortunately the gas explosion models available today are by no means perfect. The models are occasionally off with a factor two, regardless of its being empirical models or computational fluid dynamics [3]. Recently even more doubt has been cast on the models that are of use today since none of them are able to predict the damage seen at the Buncefield oil depot [4].

Two different EML studies have been carried out at the refinery in Lysekil, Sweden by two different brokers, for confidentiality reasons henceforth referred to as "Broker A" and "Broker B". Apart from the differences from using different modelling tools, there is also no set standard for which assumptions to base an EML study upon. Thus the previous two studies have generated very varied conclusions.

The maximum property damages estimated by Broker A and Broker B are 2 390 000 000 SEK, and 6 430 000 000 SEK, respectively. Broker B has identified five different scenarios that are more expensive on a property damage base than the highest one for Broker A. Even when the scenarios are based on the same process equipment failure the numbers differ. For instance a major breach on V2505, which is an intermediate storage tank for a mixture mainly consisting of butane would lead to a PD of either 1 470 000 000 SEK or 4 100 000 000 SEK, a difference of almost 300%.

In order to make a sound business decision a remodelling of the proposed scenarios has been conducted. The sources of difference between the previous models have been determined and some recommendations and thoughts on the concept of EML are given.

AIM OF THE STUDY

The main objectives of this study are as follows:

- to compare EML studies carried out by two different insurance brokers for a Swedish refinery
- to remodel and remove some of the uncertain parameters by using a third party software
- to review the EML concept and highlight areas with uncertainties that needs improvements in the future. This paper is prepared on the basis of a Master of Science thesis carried out at Chalmers University of Technology [6].

METHODOLOGY

A definition of the EML concept was given by Canaway [7] "The effect of spillage of flammable substance or inventory from the largest discrete circuit and so forth. In the EML analysis, no prediction of the ignition source location may be made in order to reduce the damage level."

In this study an EML is defined as a single release of inventory from a vessel and the resulting formation of a drifting vapour cloud. An ignition following the formation of the vapour cloud, generates an explosion, thus causing property damage. In general domino effects are not modelled in EML's and the same methodology with a single accident has been used in this study.

The two brokers EML calculations where studied in detail in order to identify and determine the sources of deviations. Thereafter, using software developed by TNO, called EFFECTS, where all physical models are described in the yellow book [8], a remodelling of all scenarios investigated by both of the companies has been done. Figure 1 summarizes the steps taken.

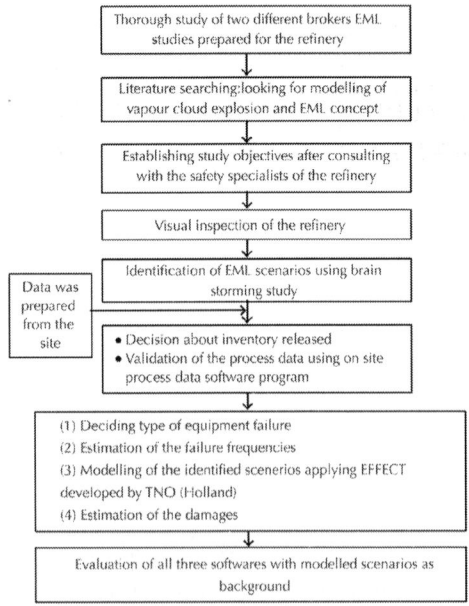

Figure 1: Methodology of the study.

The two most "expensive" scenarios presented by the two companies have also been remodelled to find out which of the two that are to be considered as the more reliable one. During the modelling a more thorough method has been used than the one employed by the two companies. Since EFFECTS allow linking of scenarios it has been possible to start with a pipe connection failure, proceed with a spray release, then model the dispersion of the gas cloud and thereafter model the overpressure after ignition. As precise input data as possible have been used, including height of release, and normal filling degree of vessels. Where different types of distillation columns have been selected as points of origin a more accurate calculation of release rates has been used.

A less improbable version of a containment failure has been used in all the modelling conducted. The bottom flange on each studied vessel has been modelled as ruptured, and the resulting jets release rate calculated. The brokers models instead assume a sudden burst, where the whole inventory all at once appear on the outside of the vessel. Though the probability of catastrophic failure of a vessel and total rupture of the piping connection modelled herein is not that different, 2 and 5 [cpm] (failure frequency of 10^{-6} per year = cpm), respectively, [9], the rupture pipe model allows for a more accurate modelling of the following step, dispersion of the cloud.

Although of interest for the operator, the concept of BI has been neglected in this study. BI is a consequence of a major accident and can be deducted from the damages from that accident. Instead of guessing delivery times for different process equipment and time for investigation and possible reengineering of the process, the BI concept has been left to those that are more suited to make such estimations.

MODELLING VAPOUR CLOUD EXPLOSIONS

Broker A is using the SLAM software, which is based on the Congestion Assessment Method. Broker B is using the ExTool software, which is based on the TNT equivalency method. EFFECTS is based on the Multi-Energy Method.

A brief description of the three different empirical methods mentioned will be given as well as the reason for not using CFD, the most accurate modelling technique.

TNT Equivalency Method (ExTool)

The TNT equivalency method assumes that a vapour cloud explosion is similar to an explosion of a high charge explosive, TNT.

A pressure-distance curve yields the peak pressure, where the distance is scaled with a TNT mass equivalent. The TNT equivalent W_{TNT} is obtained as the product between the explosion yield and the mass of hydrocarbons W_{HC} in the vapour cloud in accordance with (1):

$$W_{TNT} = 10 \cdot \eta \cdot W_{HC} \, [\text{kg TNT}]. \tag{1}$$

is the empirical yield factor, normally set between 0.03–0.05. The factor 10 is used since most hydrocarbons have a 10 times higher heat of combustion than TNT. The quota between the different heat of combustion (hc_{fuel}/hc_{TNT}) can be used for other fuel types. The main weaknesses of the TNT method is that the yield factor and pressure-distance curve are based on empirical data and not theoretically proven. Also, since TNT is a solid state explosive the difference in physical behaviour between TNT and gas explosions are substantial. With this method the predicted overpressure difference between the model and a real VCE is most pronounced close to, and far away, from to the centre of explosion. The method has a weak theoretical basis, but is used because it is simple and under most circumstances gives a reliable upper estimate [3].

Multi-Energy Concept (EFFECTS)

The multi-energy concept assumes that only the confined or obstructed part of a vapour cloud give a rise in overpressure [11]. A combustion-energy scaled distance R_{ce} is related to the distance from the explosion centre R_0 according to:

$$R_{ce} = \frac{R_0}{(E/P_0)^{1/m}} \, [\text{m}]. \tag{2}$$

P_0 is the atmospheric pressure and E is the total amount of combustion energy. E is calculated as the product of combustion energy per volume times the congested cloud volume V_{cloud}. Since the total amount of combustion energy for a stoichiometric hydrocarbon-air mixture is relatively constant regardless of the type of hydrocarbon, it is common to estimate the combustion energy E according to:

$$E \approx 3.5 V_{cloud} \; [MJ] \quad V_{cloud} \; in \; [m^3].$$
(3)

Data from explosion experiments have been fitted to the parameter R_{ce} and the overpressure for different charge strengths, dependant on, for example, strength of the ignition source and level of congestion. The charge strength is given a value in the range of one to ten, where ten represent a detonation.

Setting the charge strength and the total combustion energy is the main sources of uncertainty in the multi-energy concept.

Congestion Assessment Method (SLAM)

An assessment of the congested region is first done in order to get a reference pressure P_{ref}, which is an estimation of the maximum overpressure generated by a deflagration of a vapour cloud of propane [12]. The reference pressure is estimated with a decision tree that first takes confinement into account, then congestion or obstacles in the confined area and last whether there are strong ignition sources. There are some similarities between the choice of charge strength in Multi-Energy method and the choice of P_{ref} in CAM. If the vapour cloud does not consist of propane a fuel factor is multiplied to the reference pressure to get a maximum source pressure. With the maximum source pressure, the overpressure at a specific distance can be given by fitted data. CAM uses data from the MERGE (Modelling and Experimental Research into Gas Explosions) project.

CFD Models (Computational Fluid Dynamics)

A number of different CFD models are available today but one has to be aware of their limitations since the models are by no means perfect, even for simple geometries. MERGE was an EU-founded project that tried to determine the accuracy of some explosion models. A cubodial pipe array, shown in Figure 2, was filled with gas and thereafter ignited in the centre. The results, shown in Figure 3, depict a considerable spread even for such a simple geometry.

If one were to use a CFD model to predict the damage within a refinery it would not only take a large amount of time for the actual modelling. First of all a three-dimensional model of the refinery is needed. Further, the accuracy of the model would be lowered by the fact that a normal desktop computer today is unable to make the mesh fine enough. If one was to apply a mesh to a typical refinery area the length of each finite volume would be too large to yield an answer that is precise enough to warrant the amount of time and work needed for the modelling.

After an extensive study which included 27 large-scale experiments Ledin (1997) [3] made the following conclusion "My interpretation of the outcome of JIP-2 is that confidence can be attached to the model predictions only if the new geometry strongly resembles one of the two geometries in the database. It must be emphasised that even with the use of what appears to be in principle a more advanced model, that is, CFD-based, outside its area of validation/calibration it may in fact give little overall reduction in uncertainties over the use of simpler modelling approaches."

The obstacle geometries of a standard refinery are thus too complex to be handled by the available CFD models. Thus to screen for EML-scenarios a simpler model type should be used.

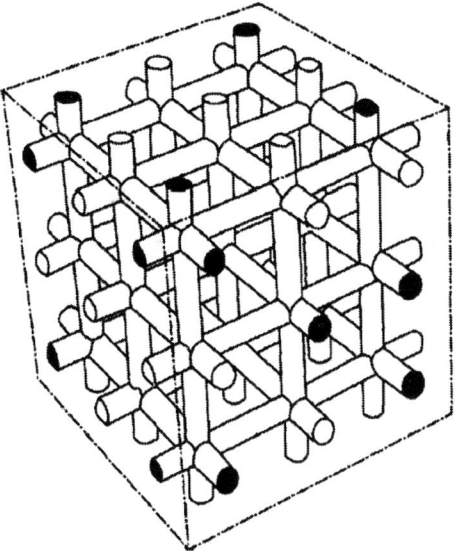

Figure 2: The cuboidal pipe array geometry used in the MERGE experiments [5].

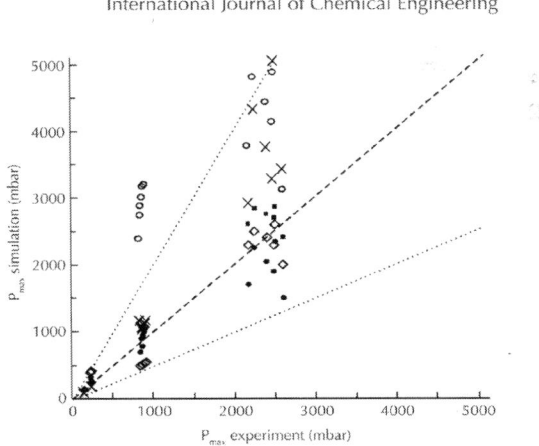

Figure 3: Comparison of experiment and simulation for MERGE large scale experiments (×) COBRA predictions, (◊) EXSIM predictions, (•) FLACS predictions, and (○) REAGAS predictions. Simulated maximum overpressure and experimental maximum overpressure [mBar], [5].

Property Damage Cost Estimation

Each process unit has an estimated total cost for rebuilding and each overpressure is associated with a specific damage percentage. Today it is common to use specific threshold values. For example between 150–350 mBar overpressure corresponds to 40% damage. Thus one calculates the damage on each subprocess area and thereafter sum up to reach the total damage cost. In reality damage is related both to duration of the overpressure and to the specific geometry of the structure [13]. Also one should consider mechanical properties of the structure, reflection, and so forth. However, calculations are usually only considering peak overpressure and positive impulse [14].

ANALYSIS

The first section presents the costs for five scenarios modelled by the two brokers. The second section discusses where in the two models the sources of difference originate. Focus will be on damage due to overpressure, overpressure decay, releasable inventory, cloud weight, and cloud drifting. The third section shows results from our study, using a third software, EFFECTS. Modelling has been made on all scenarios previously studied by the brokers. But details are given just for two scenarios, D1538 (Drum) and T2302 (Tower), due to limited space of this paper.

Section One

The data used by Broker A and Broker B in their modelling is given in Table 1.

Table 1: Data given by the client to Broker A and Broker B to use in their modelling

Equipment	Service	V [m³]	Operating V [m³]	Material	T [C]	P [Bar]	D [m]	L [m]	N of trays	Bottom V [m³]	Tray V [m³]
V2313	STRIPPER RECEIVER	43	22	C2-C5	35	10	NA	NA	NA	NA	NA
D1538	COMPRESSOR SUCTION DRUM	30,2	15	C3 = 100%	15	7,7	NA	NA	NA	NA	NA
V2505	FEED SURGE DRUM	47	24	C2-C5	35	15	NA	NA	NA	NA	NA
T2302	STRIPPER	564	83,8	C3-C10	196	10	4,87	30,3	30	55,9	27,9
T2304	STABILIZER TOWER	119	23,9	C3	160	12	2,6	22,5	30	15,9	8

V: Maximum capacity, Operating V: Estimated Operating Volume, Material: Compound composition within equipment, T: Temperature, P: Pressure, D: Diameter, L: Length, N of trays: Number of trays in separation columns, Bottom V: Estimated volume on bottom of towers, Tray V: Sum of Estimated volume on each tray.

The operating volume in the towers was estimated according to two assumptions. The bottom level was assumed to be 3 m high and the height of the liquid above each tray was assumed to be 0.05 m. For Vessels and Drums the assumption was that the liquid inventory was 50% of the total volume of the vessel.

V stands for Vessel which in these cases are cylinders lying down, D stands for Drums which are cylinders standing up and T represents Tower.

Every scenario has its origin and ignition point within the process area of the refinery. Larger vessels exist outside of the process area, for example, within the storage area. But the cost of damage to the process equipment widely outweighs the cost of damage to the storage equipment.

The estimated property damage costs (million SEK) for all scenarios are presented in Table 2.

Table 2: Summary of property damage (million SEK)

Scenario	Equipment	SLAM	ExTool Direct	**ExTool Drifted**
1	V2313	2260	3360	3710
2	D1538	2660	NA	NA
3	T2302	NA	6430	6800
4	T2304	1915	2700	4130
5	V2505	1636	3400	4100

Section Two

A number of modelling parameters could be the reason for the difference between ExTool and SLAM. Among them five were identified as the most critical potential sources for the difference. These are damage threshold values, overpressure decay, releasable inventory, cloud weight calculations, and allowed cloud drift.

Damage Thresholds

SLAM and ExTool employ two different sets of threshold values to calculate the damage percentage on process equipment. The values are shown as curves in Figure 4 for clarity reasons but are used as threshold values within the actual programs. For example, the whole area affected by an overpressure between 138–345 mar (2–5 PSI) will be 40% damaged according to ExTool.

Since damage percentage and subsequent cost depends on the overpressure as well as ignition point it is impossible to say exactly how big impact the different set of threshold values give rise to. One thing can be said though, while considering large explosions ExTool threshold values gives rise to higher costs.

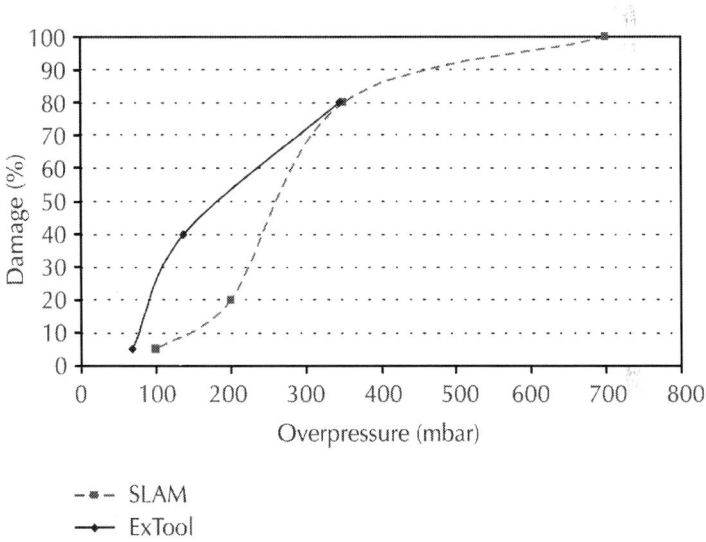

Figure 4: Comparison of threshold values

Overpressure Decay

Modelling of the pressure decay has been made with the Matlab software. In both scenarios 100 kmol of gas was used. For the typical alkane propane a yield factor of 6% has been used within the TNT model. 4% for a straight alkane raised by 2 percentage points for confinement. For propene 9% yield factor was used. Choosing 9% is done due to the fact that within CAM propene has a 150% higher fuel factor when compared to propane [12]. Thus this gives a fairer view on the actual pressure decay within the models. Congestion is in both cases set as typical within the CAM method.

As can be seen in Figure 5, with the selected parameter values, the distance to a certain overpressure does not differ that much in the near field. However, in the far field the TNT model gives lower overpressure than the CAM model.

This implies that a scenario that uses the TNT-model for its pressure decay would in fact give lower costs. In the brokers reports ExTool yields higher costs than SLAM. Thus it is probably not the use of the TNT or CAM method for overpressure decay per se that gives rise to the differences in costs.

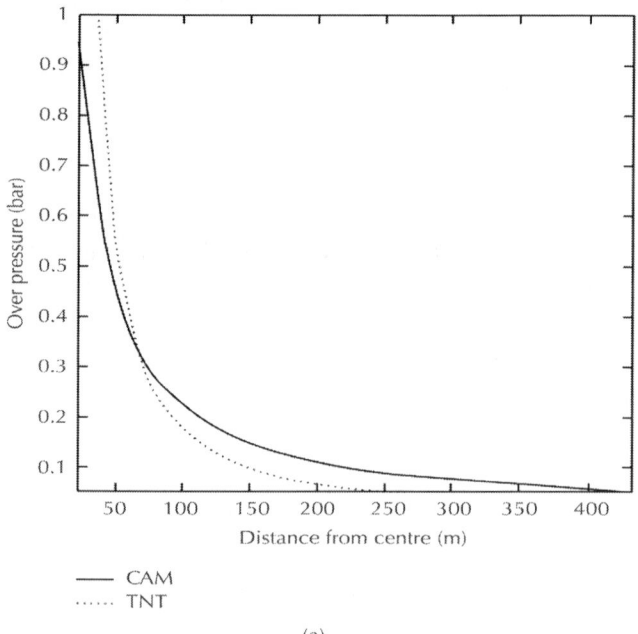

(a)

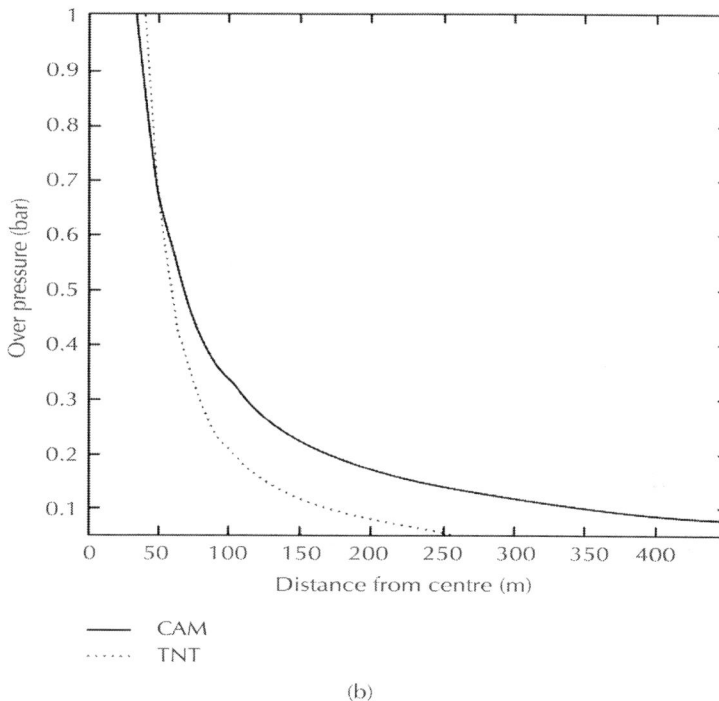

(b)

Figure 5: Overpressure to distance curves for TNT and CAM method for propane (a) and propene (b).

Releasable Inventory

For the brokers modelling the releasable amount had been set to a standard value of 50 percent of the vessel or drum size. For towers the bottom height of liquid was assumed to be 3 m and the height of the liquid on each tray was assumed to be 0.05 m.

The two brokers estimate releasable material in towers in different ways. Broker A allows only the part that is on the bottom of the tower to participate in the vapour cloud formation. According to Broker B the whole content of a tower may participate in cloud formation.

In this study a more thorough survey was conducted on the inventory of the equipment. These results are shown in Section Three.

Cloud Weights

How to calculate cloud weight is not a part of the TNT or the CAM model. But the softwares employed for the overpressure decay calculations contain a method to determine cloud weight. ExTool calculates the cloud weight as two times the flash fraction [15] [F] which in turn is calculated according to:

$$F \approx (T - T_b) \frac{c_{p,T}}{H_{\text{vap},T}}.$$

(4)

The heat capacity and heat of vaporisation are chosen at the initial temperature of the inventory. The heat capacity and heat of vaporisation depends on the temperature, hence ExTool overestimates the flash fraction. SLAM calculates the cloud weight as the flash fraction [16] and the flash fraction is calculated according to:

$$F = (T - T_b) \frac{c_{p,T\text{mean}}}{H_{\text{vap},T}}.$$

(5)

The heat of vaporisation is chosen at the boiling point for the compound at atmospheric conditions, however the heat capacity is chosen at the mean temperature between boiling temperature and the initial reference temperature. Since heat of vaporization and heat capacity both are temperature dependent the same temperature should be used for choosing physical parameters for (5). The quota gives an underestimation of the flash fraction. Also using only flash fraction to calculate cloud weight omits the entrainment phenomenon further decreasing the total cloud weight.

Cloud Drifting

ExTool has a clearly defined method to calculate maximum cloud drifting. After modelling an ignition at the point of release the cloud is allowed to travel within the 138 mBar isobar to find the position associated with the highest cost. Two objections to this method can be raised. First, since there is no connection between wind speed and dispersion, the cloud contains the same total weight no matter how far it travels. Secondly this implies that the larger the cloud, or the more reactive, the longer it will be allowed to travel before ignition. For SLAM no exact data on cloud drifting has been found. But it seems that the centre of ignition normally is within 75 m of the release point. However, so-called "engineering judgement" has been used to override the initial ignition point in one of the cases D1538. This can be done by the user if a reasonable additional drift will induce significant rise in cost.

It is not reasonable to think that a major part of the differences in damage costs could be attributed to these small differences in cloud drift allowance. ExTool scans large part of the refinery and SLAM is overridden if the cost is maximized outside the initial iteration zone.

Section Three

Using the software EFFECTS instead of SLAM or ExTool eliminates two of the parameters mentioned above, cloud drifting and cloud weight.

ExTool and SLAM use (4) and (5), respectively, to calculate cloud weight. This crude method is surpassed by EFFECTS use of so-called coupled models. The main advantage of EFFECTS is the ability to model a chain of events each with its specific method and then feeding the result into the following model. The scenarios have been modelled in the following fashion: TPDIS (bottom venting)→Spray Release→Dense Gas Dispersion→Dense Gas Explosive Mass. This means that the results from each submodel have been fed into the next to arrive at the point of interest, cloud weight. ExTool use the TNT model for its overpressure generation and decay and SLAM uses the CAM method to determine the centre overpressure and subsequent decay after ignition. In EFFECTS the coupling of the models is continued by linking an explosion model based on the Multi-Energy concept to the dense gas explosive mass. For more information on the three softwares see the yellow book [8] for EFFECTS, the ExTool theory manual [15], and the guidance document for SLAM [17].

Each scenario has its starting point in a complete rupture of the nearest flange on the bottom pipe of the specific process equipment.

For cloud weights calculations in EFFECTS four parameters are considered as key factors, due to their high impact on the end result of the cloud weight modelling. These four parameters are presented in Table 3.

Table 3: The table shows four key parameters for cloud weight modelling in EFFECTS. Hole rounding represents the edges of the hole within TPDIS. Z_0 is surface roughness, given in accordance to Hanna [18] which gives the range 0.9–1.3 for a typical refinery. Pasquill Stability F means extremely stable weather. Wind speed is self-explanatory

Parameter	Value	Unit
Hole Rounding	0.62	—
Z_0	1	M
Pasquill Stability Class	F	—
Wind Speed	1.5	m/s

The scenarios have been calculated for different stability classes and wind speeds. However Stability class F and wind speed 1.5 [m/s] were found to be the worst circumstance for every case. It should be noted that wind speeds below 1.5 [m/s] has not been tested since the dispersion model is not considered valid for such low wind speeds.

The cloud weights calculated by three different tools are shown in Table 4. These values are affected to a great extent by the releasable inventory presented in Table 5.

Table 4: Explosive mass in clouds [t] calculated by three different tools

Equipment	Broker A	Broker B	EFFECTS
V2313	4	7,7	9,1
D1538	3	NA	0,9
T2302	NA	34	28,3
T2304	7	9,5	5,2
V2505	4	9,2	9,3

Table 5: Releasable inventory [m³] used as starting points for modelling

Equipment	Broker A	Broker B	R e a l amount
V2313	22	22	20,5
D1538	15	NA	10,65
T2302	NA	83,8	52,62
T2304	16	23,9	12,6
V2505	24	24	25,2

Further elimination of uncertainty was made by using process data and drawings of the process equipment to get an enhanced certainty for the parameter releasable material. For releasable material within towers, the amount of liquid that can pass through one tray per second is lower than the decay in explosive mass due to dispersion. Therefore only the bottom content has been considered as taking part in the release to estimate the maximum explosive mass. As shown in Table 5 the difference between estimated amount and the real amount is significant.

Using EFFECTS also eliminates the Cloud Drift parameter since it is possible to see the explosive mass at each given time unit and thus eliminate extremely long drift if additional time severely impacts explosive mass.

For modelling of the overpressure obstruction has been considered as low, ignition source anticipated as high, and parallel confinement has been deemed as existing. According to Kinsella [19] these assumptions give blast strengths ranging from 5–7. In accordance with the EML concept the highest blast strength (7) in the range has been chosen.

A comparison of blast prediction models for vapour cloud explosion done in 2001 at the NRC shows how data from different models are fitted to the observed pressure data from the Flixborough accident and the accident in La Mede [20]. The close fit in this comparison for Multi-Energy, blast strength 7, indicates that this is a valid choice.

As for the threshold value parameter there is no consensus as to the usage of threshold values. It is also beyond the scope of this study to further investigate such values. However, as can be seen in Figure 6 the values used by ExTool appear to be quite conservative. All scenarios studied in EFFECTS have been modelled with both sets of threshold values to show the difference between the results. With the limited amount of data acquired it would be presumptuous to point at either as definitively correct or incorrect.

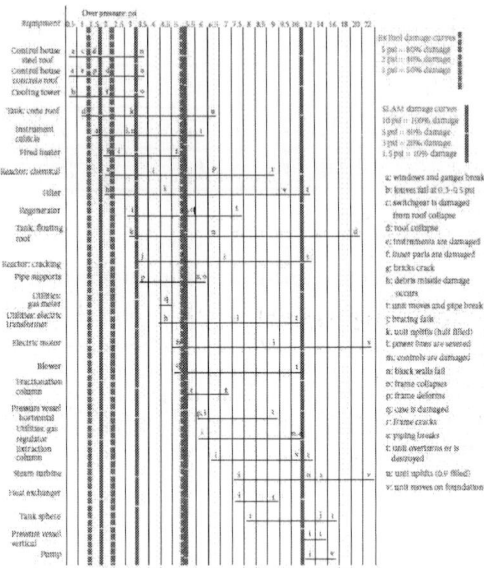

Figure 6: Threshold values from US Department of interior, Office of Oil and Gas [10].

Scenario 1: D1538

Drum 1538 contains 100% propene and is situated about 6 [m] above the ground. Average values for pressure inside temperature and volume has been taken from the process data average ranging from 2007-11-27 to 2007- 12-27. Data is thus set to P = 8.8 [bar], T_{bulk} = 14.0 [∘C], V_1 = 10.65 [m³]. To estimate a worst case scenario a total rupture of a 6″ pipe situated 6.05 [m] above ground has been simulated by EFFECTS.

The digitized cloud shape 17 s after the release represent maximum explosive mass is shown in Figure 7. The cloud drifted approximately 200 [m], which is considered far but not unreasonable. The time for the digitalization has been taken from the graph shown in Figure 8 and the total area of the cloud has been taken from the graph shown in Figure 9.

The total area of the cloud is 3050 [m²] corresponding to 900 [kg] explosive mass. The confined area was approximated at an onsite inspection, between the dispersion and explosion step, to be 965 [m²]. Figure 10shows the different damage zones using the thresholds from SLAM. Figure 11 shows the damage zones using the threshold values from ExTool.

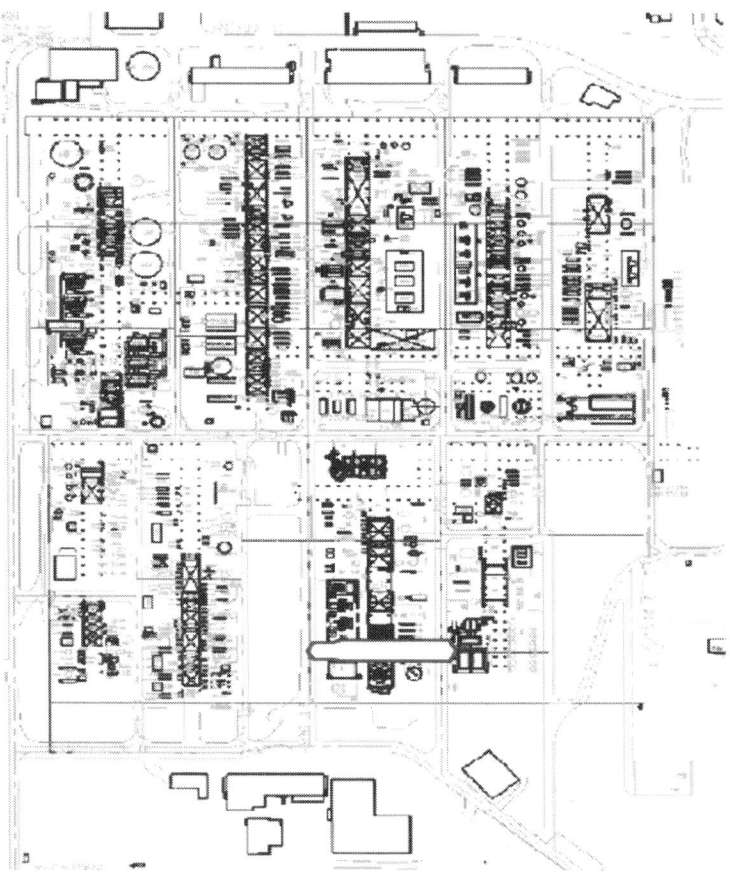

Figure 7: Cloud shape D1538. The cloud covers parts of the third and forth southern process areas.

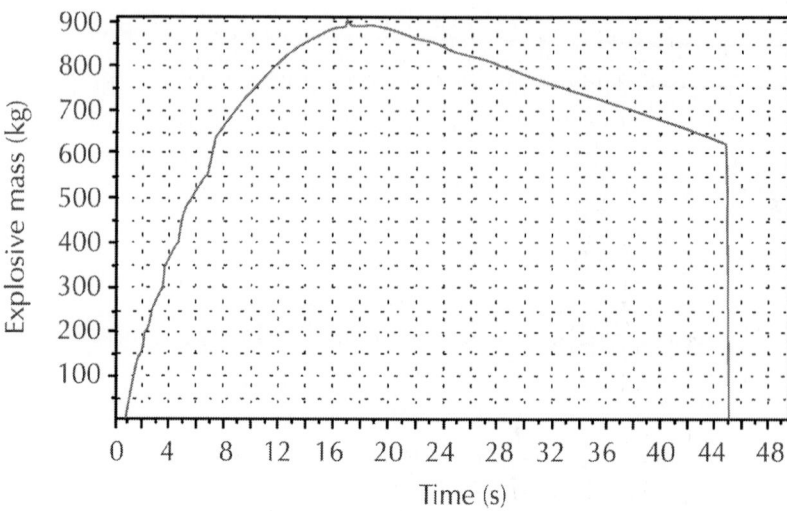

Figure 8: Explosive mass D1538.

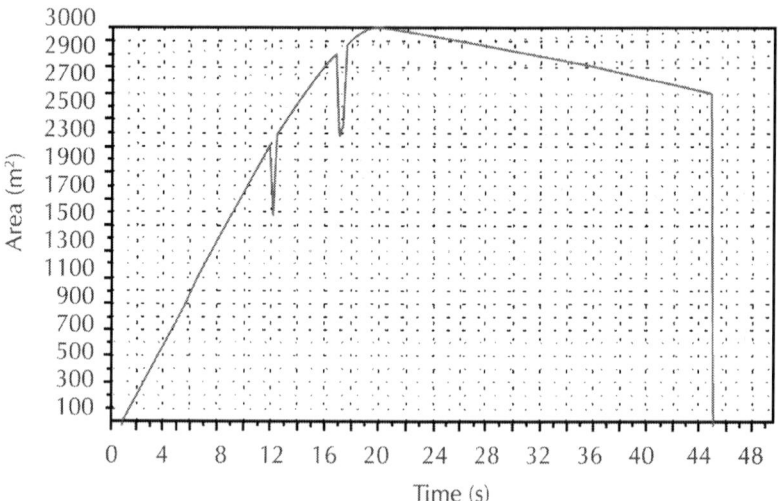

Figure 9: Area of cloud D1538.

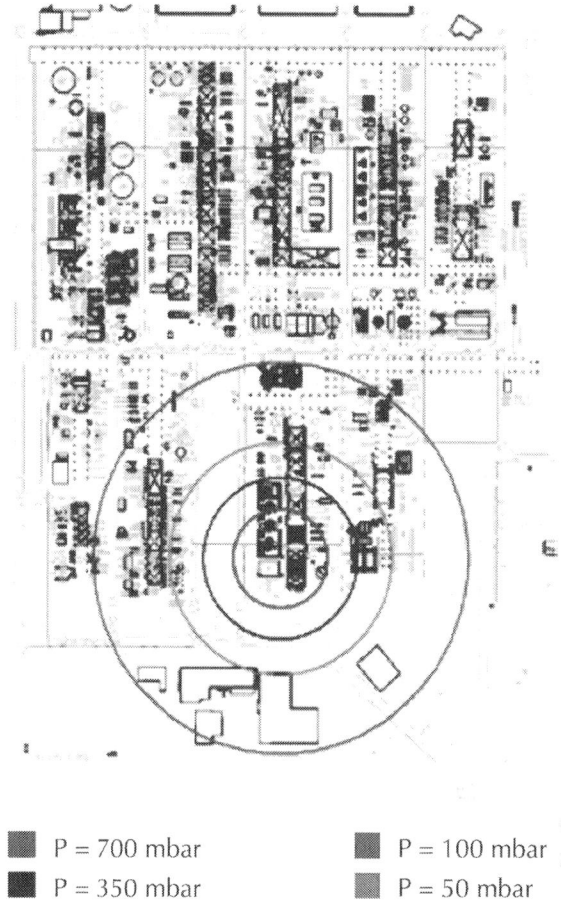

■ P = 700 mbar   ■ P = 100 mbar
■ P = 350 mbar   ■ P = 50 mbar

Figure 10: Structural damage using Slam threshold values for scenario 1: D1538.

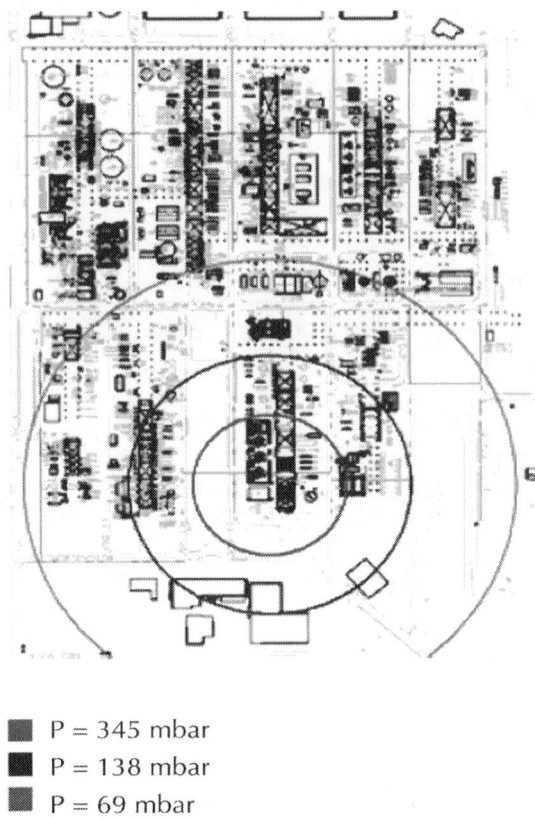

■ P = 345 mbar
■ P = 138 mbar
■ P = 69 mbar

Figure 11: Structural damage using ExTool threshold values for scenario 1: D1538.

Scenario 2: T2302

Tower 2302 main contents is C5+ [79 w%]. The temperature within the vessel has been chosen as [°C]. This is not a true process value but since EFFECTS only can handle single component releases the temperature was fitted to the initial pressure within the vessel P=11 bar. The volume that is able to participate in the cloud formation V=52.64 [m³]. The release is modelled as coming from a 20″ pipe situated 5.3 [m] above the ground. The inventory was modelled as pure Heptane.

The digitized cloud shape at 11 [s] after the release, representing maximum explosive mass, is shown in Figure 12. Within the picture it can be seen that the cloud has spread out in a both west and eastward way from the release point. This is not considered as impossible since it is one of the inherent traits of a denser than air gas to move both up- and downwind of a release point. Since the cloud has spread into two separate process areas two simultaneous explosions are modelled. The distance between these two process areas is larger than the critical separation distance for the multi-energy method [21]. The time for the digitalization has been taken from Figure 13 and the total area of the cloud has been taken from Figure 14.

The total area of the cloud is 14803 [m²] and the total explosive mass is 28300 [kg].

The confined area was approximated at an onsite inspection, between the dispersion and explosion step, the two different areas are 3531 and 1010 [m²], respectively. Figure 15 shows the different damage zones using the thresholds from SLAM. Figure 16 shows the damage zones using the threshold values from ExTool.

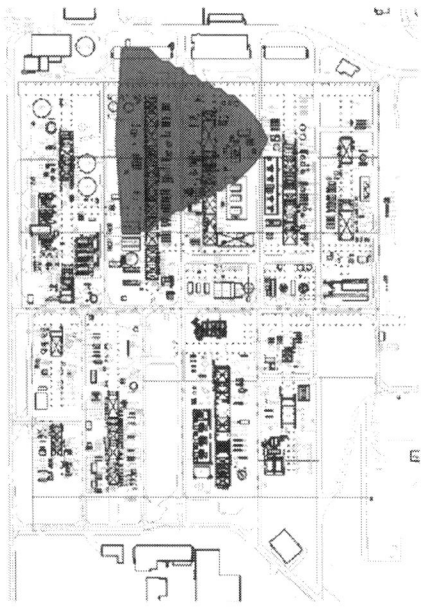

Figure 12: Cloud shape T2302. The cloud stretches over two separate process areas.

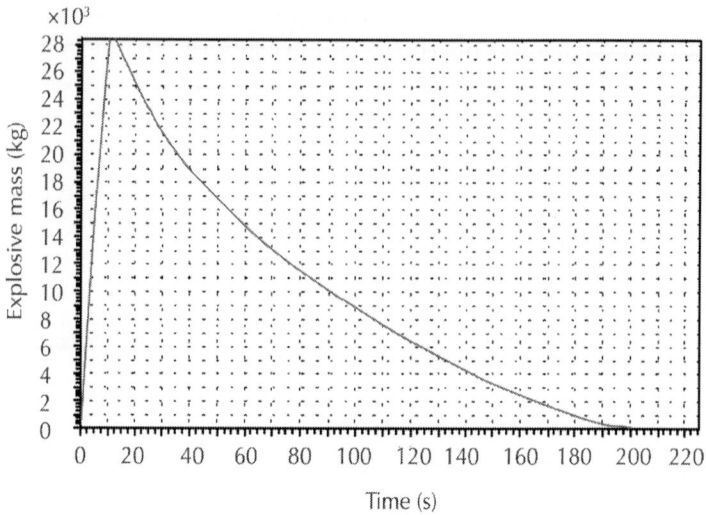

Figure 13: Explosive mass T2302.

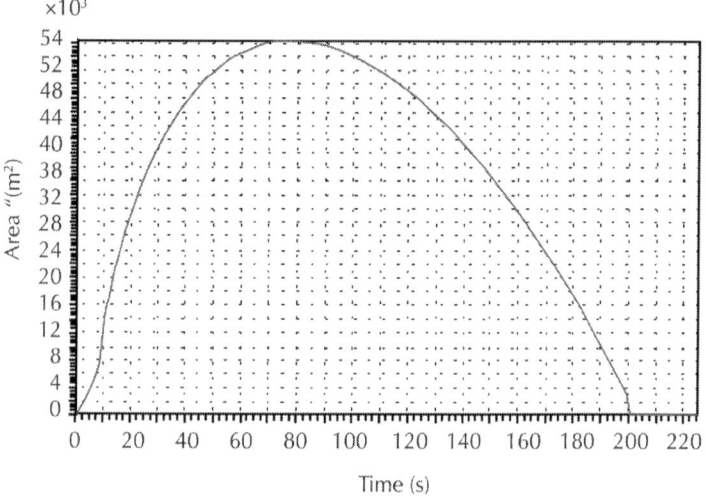

Figure 14: Area of cloud T2302.

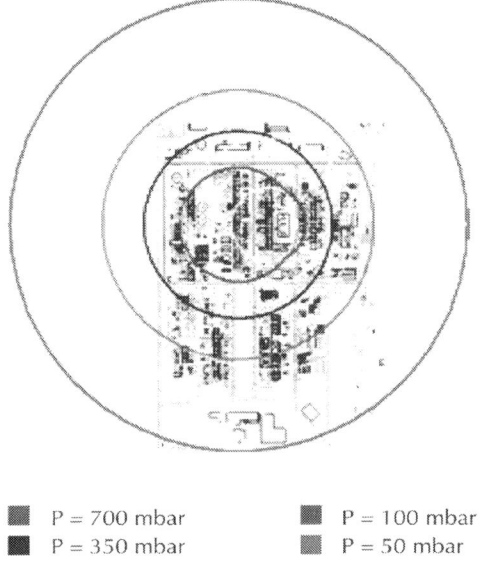

■ P = 700 mbar   ■ P = 100 mbar
■ P = 350 mbar   ■ P = 50 mbar

Figure 15: Structural damage using SLAM threshold values for scenario 2:T2302.

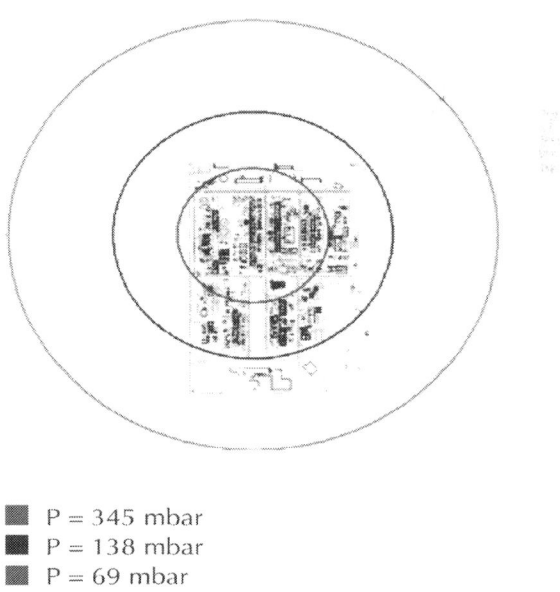

■ P = 345 mbar
■ P = 138 mbar
■ P = 69 mbar

Figure 16: Structural damage using ExTool threshold values for scenario 2:T2302.

SUMMARY OF THE RESULTS

Since there is no mechanism in EFFECTS that allows calculation of cost an alternate way for cost estimations was employed. Using Adobe Photoshop CS2 the number of pixels in each damage zone was counted as the percentage of each process unit within each damage zone.

Two apparent trends can be seen in Table 6. Except for D-1538, this study gave higher damage costs than those calculated by both Broker A and Broker B. In this study higher costs are predicted when ExTool threshold values are used. It has been previously mentioned that the ExTool threshold values seem to be on the conservative level. However this is perhaps done in order to account for the steeper pressure decay for TNT as compared to gas explosions.

Table 6: A summary of all results from this study

	Broker A	Broker B Drifted	EFFECTS: Slam TV	EFFECTS: ExTool TV
V2313				
Cost (million SEK)	2260	3710	5657	6714
Releasable inventory (ton)	22	22	22,5	22,5
Cloud weight (ton)	4	7,7	9,1	9,1
D1538				
Cost (million SEK)	2660	NA	992	1480
Releasable inventory (ton)	15	NA	10,65	10,65
Cloud weight (ton)	3	NA	0,9	0,9
T2302				
Cost (million SEK)	NA	6800	8408	9020
Releasable inventory (ton)	NA	83,8	52,62	52,62
Cloud weight (ton)	NA	34	28,3	28,3
T2304				
Cost (million SEK)	1915	4130	5251	6573
Releasable inventory (ton)	16	23,9	12,6	12,6
Cloud weight (ton)	7	9,5	5,2	5,2
V2505				
Cost (million SEK)	1636	4100	4446	5258

Releasable inventory (ton)	24	24	25,2	25,2
Cloud weight (ton)	4	9,2	9,3	9,3

It is also apparent that in all the scenarios studied the final cost suggested from EFFECTS is more in accordance with ExTool than with SLAM. The percentage of the releasable inventory that is turned into an explosive cloud is also most coherent between ExTool and EFFECTS. This suggests that the simplification to use two times the flash fraction to account for the entrainment effects is acceptable.

As can be seen from Table 6 there are large deviations in releasable inventory between scenarios modelled in EFFECTS as compared to SLAM and ExTool. Most of these differences could have been avoided if a more thorough search of the refinery inventory had been done from the start.

DISCUSSION

As a starting point for this study a definition of an EML scenario was needed. Although a written down definition was finally found which could cover the work about to be undertaken as well as the brokers previous studies this was not the only definition found. A number of different abbreviations can be found within the literature, PML (probable maximum loss), MCL (maximum credible loss), MFL (maximum foreseeable loss), EML (estimated maximum loss), and NML (normal maximum loss). All of these different abbreviations, which more or less imply the same thing, come with its own set of probability interval. However, the intervals differ between different sources as well as the definitions. Thus even choosing a proper probability level for an EML is not as easy as it might sound. The way that the two brokers handle the release from towers is clearly mirroring this lack of clearcut definition. Not only does an investigator have to find a proper definition and a proper probability interval. The investigator also has to choose one definitive source to use for probabilities of accidents.

For a refinery the size that we have studied, 140 000 m², it is unlikely that the modelling of EML scenarios actually helps the decision process. With clever use of the models it is possible to let most of the scenarios vary between denting the closest equipment to total annihilation of the whole process area. This kind of modelling might not have any use until no matter how far you stretch the model there are still parts of the refinery outside of the blast radius. Although one should remember that domino effects are neglected in EML modelling.

In order to choose a model one must consider the purpose of the modelling as well as the necessary precision of the model. In Table 7 a set of criteria is listed in order to help with such a decision. The criteria have been defined in accordance to Transparency: the ease of finding, and interpreting how a certain model works;Input Demand: The time that needs to be spent in order to collect the necessary data for using the model;Complexity: The level of knowledge needed to use the model which also reflects the amount of influence the choice of analyst might have; Precision: A model ability to accurately reflect the reality of a VCE.

Table 7: Model choice criteria

	Transparency	Input demand	Complexity	Precision
ExTool	High	Low	Low	Low
SLAM	Medium	Medium	Medium	Low
EFFECTS	Medium	High	High	Medium

As EML is so ill defined and the number of uncertainties in the dispersion, drift, ignition, discharge rates, damage thresholds, equipment failures and so forth, is so high there is no reason to aim for a time-consuming, although somewhat more accurate model. As a primary choice for EML modelling, if at all such modelling must be conducted, ExTool stand out with its ease of interpretation owing to its high openness and low complexity. In addition the low amount of time needed to collect the data further strengthens its position. Also, explosion modelling is a specialist field and it is highly inappropriate to mask the very high insecurity in the modelling behind complex models. Although, no matter the choice of model an increased openness about their limitations would be in its place.

When comparing the models one can see that the energy released as overpressure can be matched setting parameters such as the confinement percentage and the yield coefficient at the right levels. In the results reported herein we see the effect of these two parameters. The energy released as overpressure has been higher for the Multi-

Energy method, used by EFFECTS, simply because we have deemed the confined area as larger than the corresponding yield coefficient used in ExTool (TNT model) by Broker B. One remark concerning the three methods dissected here must thus be made. Any of the three methods can be made to fit most historical cases. In the TNT method the analyst can adjust the yield, in SLAM (CAM method). can be adjusted and in EFFECTS (Multi-Energy method) the charge strength and confinement values can be adjusted. All these three overpressure models mentioned are also dependant on a sensible choice of vapour cloud weight, which in historical cases seldom can be definitively known. Hence, any claim, also the ones made within this report, on historical accuracy should be taken with a grain of salt. These flexibilities of the models are strengths as well as weaknesses. As with any software model it is almost always possible to get the response one wants. Modelling must therefore be made by a competent analyst and with a predefined set of rules that must not be broken.

CONCLUSIONS

The EML concept as it is used today is a rather loosely defined method to compute the maximal damage due to a large-scale accident. Also, any modelling at so extreme conditions as those used for EML scenarios are bound to be uncertain since the scale is balancing on the validity range of any model used. Hence improvements should be made not only to the models used but also to the EML concept itself. Clear cutoff values for the probabilities of an accident should be used to avoid the "not plausible" argument sometimes heard. As for improving the models themselves, no clear reason for working with threshold values when it comes to overpressure damage can be found. A continuous curve seems more fitting in the age of computers. The possibility to shift such a curve to account for the difference in overpressure sustainability between different types of process equipment could also be explored. Further, phenomenological models have been left out of this study but the result of such a model could prove interesting, at least as comparison.

All in all, there are many aspects to investigate further in order to make potential loss predictions more reliable, and this should be well worthwhile since much money is at stake when plant owners and insurers decide on insurance limits and premiums.

REFERENCES

1.   Marsh, The 100 Largest Losses 1972–2001, 20th edition, 2003.

2.   Marsh, Loss Control Newsletter, 1997.

3.   C. J. Lea and H. S. Ledin, "A review of the state of the art gas explosion modelling," HSL/2002/02.

4.   Explosion Mechanism Advisory Group Report; Buncefield Major Incident Investigation Board, Health & Safety Commission, 2007.

5.   N. R. Popat, C. A. Catlin, B. J. Arntzen et al., "Investigations to improve and assess the accuracy of computational fluid dynamic based explosion models," Journal of Hazardous Materials, vol. 45, no. 1, pp. 1–25, 1996.

6.   A. Lundkvist and M. Gustavsson, "Estimation of maximum loss: a comparative study".

7.   N. Gibson, Ed., Major Hazards Onshore & Offshore, Institution of Chemical Engineers Symposium Series No. 130, Taylor & Francis, Oxford, UK, 1992.

8.   CPR-14E, "Yellow Book," 2005.

9.   C. Delvosalle, C. Fiévez, and A. Pipart, Aramis Project Deliverable D.I.C WP 1, July 2004.

10.   M. M. Stephens, "Minimizing damage to refineries from Nuclear attack, natural and other disasters," Tech. Rep. AD-773048/4, US Department of interior Office of Oil and Gas, 1970.

11.   A. C. Van den Berg, "The multi-energy method. A framework for vapour cloud explosion blast prediction," Journal of Hazardous Materials, vol. 12, no. 1, pp. 1–10, 1985.

12.   J. S. Puttock, Major Hazards Onshore & Offshore, vol. 2 of Institution of Chemical Engineers Symposium Series 139, Institution of Chemical Engineers, 1995.

13.   E. Salzano and V. Cozzani, "The analysis of domino accidents triggered by vapor cloud explosions,"Reliability Engineering and System Safety, vol. 90, no. 2-3, pp. 271–284, 2005.

14. P. Schneider, "Limit states of process equipment components loaded by a blast wave," Journal of Loss Prevention in the Process Industries, vol. 10, no. 3, pp. 185–190, 1997.

15. E. Zirngast, ExTool Theory Manual, 2008.

16. Broker, A Report: Private Communication.

17. J. S. Puttock, Major Hazards Onshore & Offshore, vol. 2 of Institution of Chemical Engineers Symposium Series 139, Institution of Chemical Engineers, 1995.

18. S. R. Hanna and R. E. Britter, Wind Flow and Vapor Cloud Dispersion at Industrial and Urban Sites, 2002.

19. K. G. Kinsella, "A rapid assesment methodology for the prediction of vapour cloud explosion overpressure," in Proceedings of the International Conference and Exhibition on Safety, Health and Loss Prevention in the Oil, Chemical and Process Industries, Singapore, 1993.

20. J. Jiang, Z. G. Liu, and A. K. Kim, "Comparison of blast prediction models for vapor cloud explosion," in Proceedings of the Combustion Institute/Canada Section Spring Technical Meeting, vol. 44715, pp. 23.1–23.6, NRCC, May 2001.

21. A. C. van den Berg and N. H. A. Versloot, "The multi-energy critical separation distance," Journal of Loss Prevention in the Process Industries, vol. 16, no. 2, pp. 111–120, 2003.

2

Improving the Demulsification Process of Heavy Crude Oil Emulsion through Blending With Diluent

K. K. Salam, A. O. Alade, A. O. Arinkoola,
and A. Opawale

Petroleum Engineering Unit, Department of Chemical Engineering,
Ladoke Akintola University of Technology (LAUTECH), PMB 4000,
Ogbomoso, Nigeria

ABSTRACT

In crude oil production from brown fields or heavy oil, there is production of water in oil emulsions which can either be controlled or avoided. This emulsion resulted in an increase in viscosity which can seriously affect the production of oil from sand phase up to flow line. Failure to separate the oil and water mixture efficiently and effectively could result in problems such as overloading of surface separation equipments, increased cost of pumping wet crude, and corrosion problems. Light hydrocarbon diluent was added in varied proportions to three emulsion samples collected from three different oil fields in Niger delta, Nigeria, to enhance the demulsification of crude oil emulsion. The viscosity, total petroleum hydrocarbon, and quality of water were evaluated. The viscosity of the three emulsions considered reduced by 38, 31, and 18%. It is deduced that the increase in diluent blended with emulsion leads to a corresponding decrease in the value of viscosity. This in turn enhanced the rate of demulsification of the samples. The basic sediment and water (BS&W) of the top dry oil reduces the trace value the three samples evaluated, and with optimum value of diluent, TPH values show that the water droplets are safe for disposal and for other field uses.

INTRODUCTION

Emulsion is defined as a system in which one liquid is relatively distributed or dispersed, in the form of droplets, in another substantially immiscible liquid. Emulsions have long been of great practical interest due to their widespread occurrence in everyday life which occurs due to reliance of the behaviour of the emulsion on the magnitude and range of the surface interaction. They may be found in important areas such as food, cosmetics, pulp and paper, biological fluids, pharmaceutical, agricultural industry, and petroleum engineering. In production and flow assurance, the two commonly encountered emulsion types are water droplet dispersed in the oil phase and termed as water-in-oil emulsion (W/O) and if the oil is the dispersed phase, it is termed oil-in-water (O/W) emulsion [1].

Water-in-oil crude oil emulsions may be encountered at all stages in the petroleum production and in processing industry. With presence of water, they are typically undesirable and can result in high pumping costs and pipeline corrosions and increase the cost of transportation [2]. Reduced throughput is needed to introduce special handling equipment, contribute to plugging of gravel pack at the sand phase [3], and affect oil spill cleanup [4].

In their research work, Micheal et al. used bottle test method to simulate field condition of four emulsion samples (two Canadian and two Venezuelan emulsions) in order to determine the variables that affect emulsion stability. They were able to evaluate response to the different emulsion based on bottle test data by introducing thirty-six different demulsifiers to enable them to probe emulsion stability. Linear regression and partition tree analysis were used to analyze the effect of various variables on emulsion stability and were able to conclude that solid content significantly affects emulsion stability. Beside solid content crude oil properties, water chemistry and process condition also influence emulsion stability [5].

Christophe et al. evaluated and compared emulsion formed by different parts of the indigenous amphiphiles (the light, the intermediate, or the heavy ones) to determine their contribution to emulsion stability. The emulsions formed with the light and intermediate fractions separated immediately when the agitation stopped. The most stable emulsions were formed with the fraction of crude that distilled at temperature greater than 520°C, suggesting that the amphiphiles with the highest molecular weight, that is, resins and asphaltenes, play a major role in the protection of water droplet against coalescence [6]. This is consistent with many recent findings that the presence of these components enhanced w/o emulsion stability (Rondón et al. and Ekott and Akpabio [7, 8]). Others factors that affect emulsion stability are fine solids, temperature, size of water droplet, and brine composition [9], which is consistent with the work of previous authors [5, 6].

Despite the success of enhanced oil recovery (EOR) process, one of the problems associated with the process is emulsion problem. Efeovbokhan et al. observed that physical factors that enhance oil recovery can also greatly contribute to the formation of very stable emulsions because EOR-induced emulsions are established by surfactant/polymer (SP) and alkaline/surfactant/polymer (ASP) processes which makes breaking of emulsion different from naturally occurring emulsions which are stabilized by asphaltenes and resins [10]. Traditional demulsifiers are often not effective on emulsions created by chemical floods; therefore, the performance of demulsifier in surfactant/polymer–flooding-induced emulsion depends on the selection of the best demulsifier with respect to the system under consideration [11]. In breaking of surfactant/polymer-flooding-induced emulsion with the use of surfactant, Oseghale et al. worked on separation of oil-water emulsions expected during chemical enhanced recovery operations using crude oil from a field in Niger delta during surfactant/polymer flooding operation. Surfactant N-octyltrimethyammonium bromide (C_8TAB) was used as the demulsifier and a dosage between 200 and

300 ppm was the optimum dose that yielded oil and water phases with oil content reduction from 550 to 70 ppm after 4 h. Microscopy test confirmed that addition of N-octyltrimethyammonium bromide (C_8TAB) produced significant coalescence shortly after it was added to the emulsion, which is in agreement with an increase of the oil droplet size in the presence of the demulsifier. Their findings show that this investigation worked with the principles of using cationic surfactants as demulsifier [12].

With various problems encountered with the presence of emulsion in our system, there is need to find ways of controlling existence of emulsion or preventing it from forming in our system. One of the ways of controlling problems encountered by crude oil emulsion is the ability to predict crude oil behaviour both at the sand phase and during production by building a robust predictive model [13]. Emulsion formation or break up either for oil in water or water in oil emulsion can be characterized based on the property and type of crude oil involved in the formation or break up of emulsion which can assist in formulating method of preventing formation of such emulsion [14]. Nuraini et al. selected four groups of demulsifiers which are amine, natural, polyhydric, and alcohol demulsifier groups serving as breaking agents of stable emulsion. Their findings show that amine demulsifier group exhibited the highest efficiency to break the emulsion compared to polyhydric, alcohol, and natural groups and that demulsifier efficiency depends on two-factor solubility of demulsifier either in water or oil and molecular weight of demulsifier [15].

It has been established from the literatures that one of the ways of breaking stable emulsion is introduction of low dose of demulsifiers. For comprehensive methods of breaking emulsion, the work of Hanapi et al. treated that aspect [2]. Micheal et al. used chemical demulsifiers and statistical analysis to classify emulsion. They obtained emulsion from the field and treated the emulsion with thirty-eight chemicals that serve as demulsifiers at nine different sites. The tests were tailored towards determination of water droplet, oil dryness, and oil-water interface which were analyzed using several statistical tools. A correlation was developed for water droplet, oil dryness and oil-water interface. The results show that water droplet significantly affect oil-water interface than oil dryness [16].

Crude oil emulsions are complex and should be characterized as completely as possible. Droplet-size distribution, interfacial phenomena, and the nature of organic and inorganic components are important. The viscosity of the emulsion is affected by both the water content and droplet size distribution [17, 18]. The increase in aqueous phase of the emulsion leads to an increase in viscosity of emulsion which in turn aggravates flow of emulsion in conduct either at the sand phase or through the surface facilities [3, 19]. Stable water-in-oil emulsions have been generally found to exhibit high interfacial viscosity and/or elasticity modulus. Viscosity of crude oil emulsion was found to increase with increase in water and decreased with increase in speed of rotation of spindle when demulsifier is added [20]. The increase of the interfacial rheological parameters has been attributed to non-Newtonian nature of emulsion [20] and physical cross-links between the asphaltene particles adsorbed at the water-oil interface [21]. Demulsification of emulsion proved to be a good method of breaking emulsions but with an influence of viscosity still unaccounted for in most of the researches; this research will study the effect of adding a diluent to emulsion samples treated with diluent for three different water in oil crude emulsions collected from three different oil fields from three operators in Niger delta, Nigeria.

MATERIALS AND METHODS

Sampling

Fresh crude oil emulsions were collected from the three oil fields flow stations operated by three different operators in the Niger delta in Nigeria, namely, Fields A, B, and C.

At the sampling points in all the three oil field mentioned above, crude oil was collected at both east and west directional sampling pipes. This to ensure that pure emulsion interface is collected and not either gas or water phase. The emulsions are collected in a tightly sealed container. The experiment was carried out after four hours from the time of sampling to avoid ageing of the crude oil. Table 1 show the initial properties of the three water-in-oil emulsion samples used for the experimental work.

Table 1: Properties of crude oil emulsions used for the analysis

Field A	Field B	Field C	
Temperature (°C)	55	60	50
Production rate (m³/day)	11,000	32,000	41,000
Viscosity (mPas⁻¹)	80	100	215
Residence time (hrs)	11	7	15
API gravity (°)	23	22	21
Water cut (%)	51	8	10
Demulsifier volume used (ppm)	3	18	6

Gasoline used as the diluent for this experiment was gotten from the Nigeria National Petroleum Cooperation (NNPC) Refinery, Port Harcourt, Nigeria. Two types of demulsifiers were used by the operators for breaking of the emulsions formed. Fields A and C used PhaseTreat 4633, while Field B used PhaseTreat 6074.

The equipment used for the analysis are water bath, Checktemp1 digital thermometer, Cannon Fenske viscometer, Model HT 5001-201, six-ounce Pyrex bottles with volume 100 mL, Beaker (100 mL), Socorex Syringe micropipette, Model: Dossy TM 174 premium, Centrifuge Machine: Robinson Centrifuge, Model T.0.2, serial no. T724, Wooden product bottle shaker, and 10 mL measuring cylinder.

Experimental Procedure

Each of the crude oil samples was analyzed in different setups; the three crude oil samples were treated according to the properties of the oilfield where they were collected. These properties vary in terms of temperature, rate of chemical injection, nature of process terminal, and time of processing, which will dictate the type of demulsifier chemical to be used. A water bath was set up and maintained at a temperature of 60°C equivalent to the average process temperature of the oil fields. This temperature was held constant to neglect the effect of temperature on the viscosity of the crude oil samples.

Six test bottles of capacity 100 mL were labeled according to their corresponding wells with A, B, and C, with suffixes 1 to 6 on each of the wells. The suffix 1 denotes 0 mL of diluent, 2 is 2 mL, 3 is 4 mL, 4 is 6 mL, 5 is 8 mL, and 6 is 10 mL of diluent. The bottles were filled with crude oil and gasoline to make up a volume of 100 mL. Prior to addition of diluent to the emulsion, demulsifier was added in a ratio of one-third of the amount used by the operators where the samples were collected. The samples (emulsion + gasoline) were placed in a bottle shaker and agitated thoroughly with 50 vertical shakes and 50 horizontal shakes to homogenize the diluent with the continuous phase of the emulsion. The bottle was returned to water bath after blending for ten minutes after which percentage-free water was recorded.

The viscosities reading of various combinations of the blend of demulsifier, emulsion, and diluent were obtained using the Cannon Fenske viscometer according to the procedure recommended by ASTM D445 (Norman) [22].

Basic sediment and water of the emulsion was determined using the method described in the published work of Sunil et al. [9]. Total petroleum hydrocarbon was measured by using TPH analyzer (Model HC-404). A sample of the effluent water was taken and fed into this analyzer and the reading recorded in parts per million, ppm.

RESULTS AND DISCUSSION

Flow Assurance

Effect of diluent on viscosity is illustrated in Figure 1. Gasoline was added to the three crude oil emulsions from 2 mL to 10 mL and emulsion volume is reduced from 100 to 90 mL. There is viscosity reduction when diluent is introduced. The reduction in viscosity is proportional to the increase in the volume of gasoline. Effect of viscosity reduction is a function of initial viscosity of the emulsion because from the graph the reductions of viscosity in fields A, B, and C are 38, 31, and 17%, respectively when 10 mL of gasoline was added to them and their initial viscosity are 80, 100, and 215 mPas which means that the sample with lowest value of viscosity experienced the highest percentage reduction in viscosity value and the sample with the highest viscosity experienced the lowest percentage reduction in viscosity value.

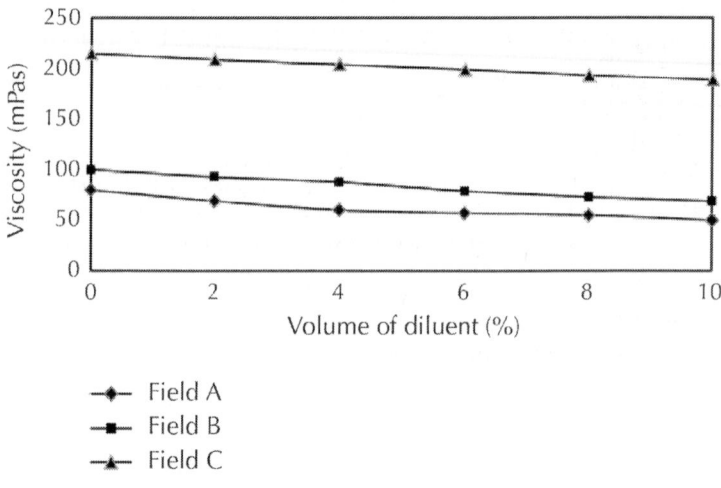

Figure 1: Volume of diluent against viscosity of emulsion.

Rate of Separation of Water

It was observed from Figures 2 and 3 that introduction of diluent affect Basic sediment and water (BS&W) of crude oil emulsion. The BS&W of Field A crude oil emulsion sample was originally 0.5% when treated with an injection rate of 1 ppm, and without blending with diluent. The value reduces as the volume of diluent increases until 8 mL when the value is zero. There is reduction in the value of BS&W of samples from fields B and C which was initially at 0.7% after the addition of 6 mL and 2 mL of demulsifiers to them to zero and 0.2% when the volume of the diluent was increased to 10 mL. Also viscosity of emulsion plays an important role in the analysis of BS&W because Field A reduces its value of BS&W to zero when the volume of diluent is 8 mL, B when the volume is 10 mL, and for C the value is at 0.2% when the volume of the diluent is at 10 mL. Field A used the lowest amount of diluent because it has the lowest viscosity while C has the the highest amount of diluent since it has the highest viscosity. Therefore, it is established that the diluent is capable of increasing reduction of BS&W in crude oil emulsions.

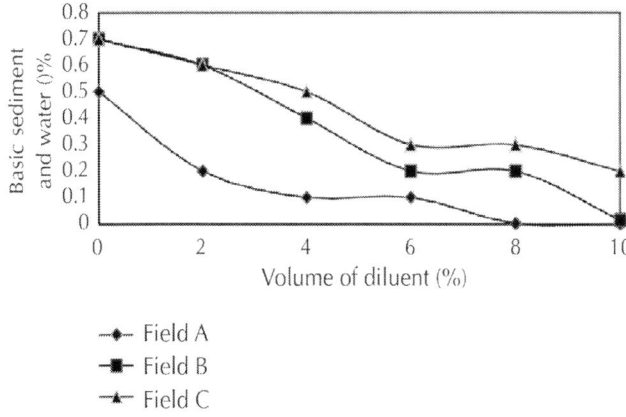

Figure 2: Volume of diluent against basic sediment and water.

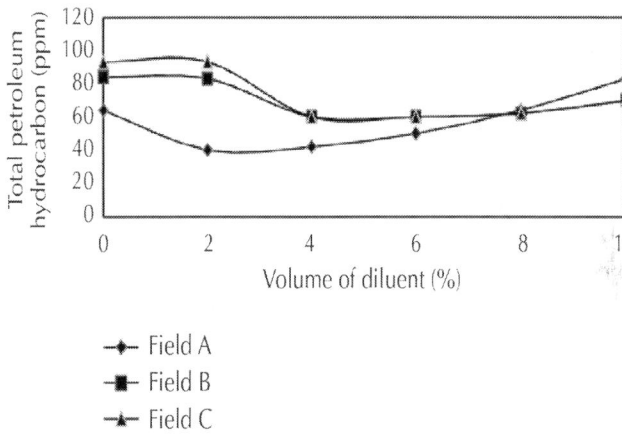

Figure 3: Volume of diluent against basic sediment and water.

Total Petroleum Hydrocarbon

The TPH of each sample initially reduces with the increase in the amount of gasoline added, but later started increasing after a particular blending point. This undesirable effect is believed to be caused by excess gasoline in the mixture finding their way into the aqueous phase.

Field A crude sample maintained a good TPH of 64 ppm which reduces as the diluent value increased between 0 and 2 mL, but there is a sharp increase in the value of TPH as the volume of diluent is increased above 2.2 mL. These reduction and rising of the TPH value with the increase in the diluent volume are attributed to the relative tightness of the crude oil emulsion. Tightness is the degree at which the water droplets are held in suspension and resist separation.

Fields B and C crude emulsions demonstrated high TPH values of 84 ppm and 93 ppm. Initially, blending showed less effect of diluent on the TPH of these two crudes between 0 and 2 mL. The TPH values of B and C reduced to 60 ppm at 4 mL of diluent, which was constant till 6 mL of diluent. Above 6 mL of diluent, there is an increase in the value of TPH fields of B and C to 70 when the volume of diluent is 10 mL. Apart from tightness of the emulsion, excess diluent can penetrate aqueous phase of the emulsion which will increase the value of its TPH.

Bottle Test

The demulsification bottle test was carried out and results on water droplet are taken after 5, 20, 30, 60, and 720 minutes. Water droplet is the separation of water from the surface of emulsion formed. The effect of addition of diluent on each crude oil sample was monitored on the rate of water droplet from each of the emulsion samples. The suffixes after the fields denotations A, B, and C indicated the variation of diluent concentration added to the emulsion samples which read 1, 2, 3, 4, 5, and 6 for 0, 2, 4, 6, 8, and 10 mL of diluent concentration.

From Figure 4, depending on the amount of diluent blended with the emulsion there was a corresponding increase in the rate of water droplet with time. When 2 mL of diluent was blended with emulsion, there was no water droplet after 5 minutes; it increased to 6% at 20 minutes and 22% at the end of 60 minutes after which there was no further droplet till the end of 720 minutes which was illustrated in A-1 in Figure 4. In A-2, the trend was similar to that of A-1 but the final water droplet value is 24%. In A-3 and A-4, there was a water droplet of 4 and 5% at 5 minutes which increased to 6 and 20% after 20 minutes, 24% at the end of 30 minutes after which there was no further droplet till the end of 720 minutes.

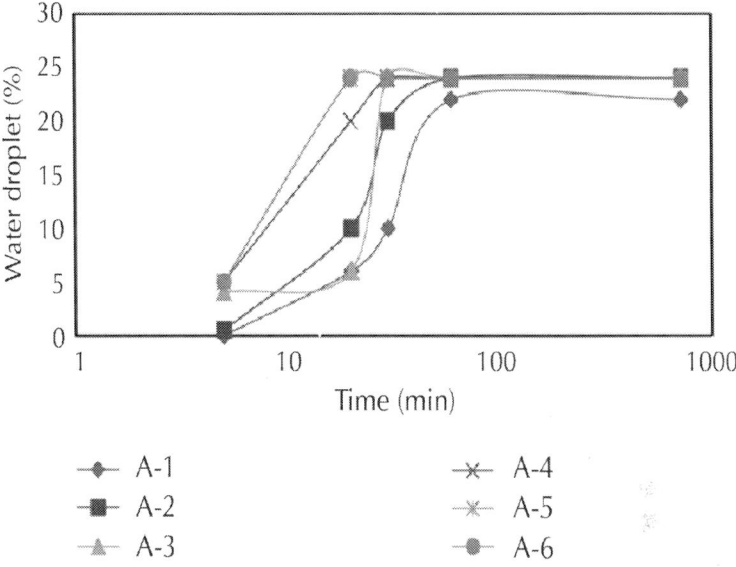

Figure 4: Water droplet against time at various emulsion/diluent ratio for Field A.

In A-5 and A-6, water droplet was 5% at 5 minutes, 24% at the end of 20 minutes, and remained constant till the end of 720 minutes. Generally, it was observed that low amount of diluent take longer time for water to drop from the emulsion but as the volume of diluent increased, the time required for water to drop out of the emulsion decreased.

Figure 5 shows the behavior of change in diluent concentration with rate of water droplet for emulsion samples collected from Field B. When no diluent was blended with the emulsion samples obtained from Field B, there was no droplet of water until after 60 minutes with a value of 4% and progressively increased to 6% at the end of 720 minutes. In B-2 to B-4, the trend of the curve followed the trend experienced in B-1 only that the rate of water droplet was faster and higher than that of B-1 with a value of 8, 10, and 10%, respectively, for B-2, B-3, and B-4 at the end of 720 minutes. In B-5 and B-6, water droplet was experienced earlier than the previous four situations with droplet of 3 and 4% at 5 minutes; it increased to 5 and 10% at 20 minutes and was 12% from 30 to 720 minutes when the analysis was terminated.

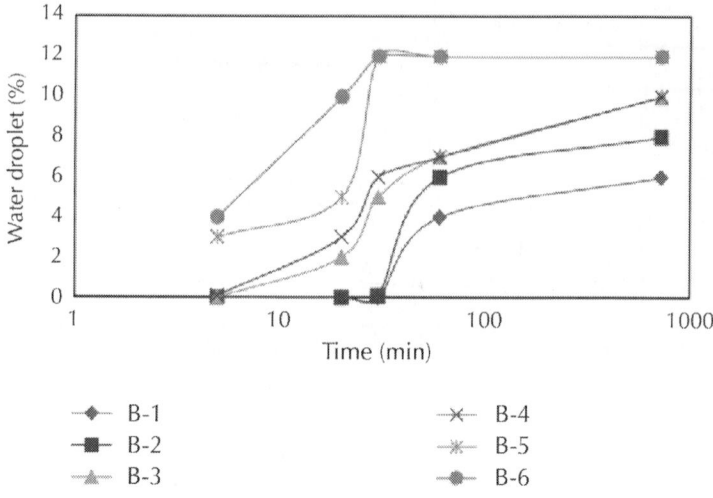

Figure 5: Water droplet against time at various emulsion/diluent ratio for Field B.

When no diluent was blended with the emulsion obtained from Field C and when 2 mL is blended with it, the behavior of their chart was similar and illustrated in C-1 and C-2 in Figure 6. There was no water droplet in the two charts until at 30 minutes with water droplet value of 0.1% which increased to 4 and 5% at 720 minutes. C-3 and C-4 followed the trend observed in C-1 and C-2 only that water droplet rate was faster with a value higher than that of C-1 and C-2. The value of water droplet at 720 minutes in C-3 and C-4 are 9 and 10%. In C-5 and B-6, water droplet was experienced earlier than the first four situation. After 5 minutes, water droplet was 1 and 2% which increased to 4 and 5% at 20 minutes. The value of water droplet remains constant at 30 till 720 minutes at 10%.

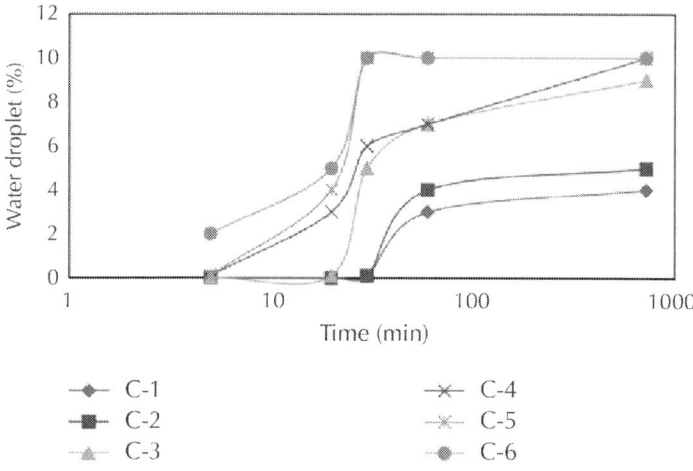

Figure 6: Water droplet against time at various emulsion/diluent ratio for Field C.

Water Quality

The quality of water droplet and observation at the oil and water interface after separation for the three crude oil emulsions are captured in Table 2. For the three crude oil emulsion samples when diluent is not blended with the emulsion, the water quality is dirty and the interface between the water droplet and oil phase is cloudy after 720 minutes. As the diluent volume blended with emulsion increased there, is an improvement in the quality of water change from dirty to clean (i.e., there is no residual emulsion or oil in the water) and the interface between oil and dropped water changes from stained to sharp (i.e., there is a distinct different between water phase and oil phase).

Table 2: Effect of diluent on water quality and interface for emulsions after 720 minutes

Volume of emulsion	Volume of diluent	Field A		Field B		Field C	
		Water quality	Interface	Water quality	Interface	Water quality	Interface
100	0	Dirty	Cloudy	Dirty	Cloudy	Dirty	Cloudy
98	2	Fair	Stained	Dirty	Stained	Dirty	Cloudy
96	4	Fair	Sharp	Fair	Stained	Dirty	Stained
94	6	Clean	Sharp	Fair	Sharp	Fair	Sharp
92	8	Clean	Sharp	Clean	Sharp	Clean	Sharp
90	10	Clean	Sharp	Clean	Sharp	Clean	Sharp

CONCLUSIONS

Generalized conclusions are hence drawn from the observation of the three samples of crude oil used for this bottle test as follows:(i) the viscosity of the three water-in-crude oil emulsions considered is inversely proportional to the increase in volume of diluent (gasoline) blended with emulsion. Also the effect of gasoline on the viscosity reduction was observed to be a function of the heaviness of the crude oil emulsion because the higher the viscosity of the emulsion the lower the reduction percentage in its viscosity value. Blending of emulsion reduced the viscosity of the three samples considered by 38, 31, and 17%, respectively.(ii)Basic sediment and water (BS&W) reduces as the volume of diluent blended with the emulsion increases. This is also a function of viscosity of emulsion prior to blending because BS&W decreased with the decrease in the value of viscosity.(iii) Total petroleum hydrocarbon (TPH) decreases with the increase in the volume of diluent until optimum concentration of diluent is reached and the TPH increases with further increase in volume of diluent. Optimum value varied for the three crude oil emulsions considered in the analysis. However, above optimum volume of diluent the TPH of the effluent increases which creates another problem when it comes to water disposal.

REFERENCES

1.   D. Langevin, S. Poteau, I. Hénaut, and J. F. Argillier, "Crude oil emulsion properties and their application to heavy oil transportation," Oil and Gas Science and Technology, vol. 59, no. 5, pp. 511–521, 2004.

2.   M. Hanapi, S. Ariffin, A. Aizan, and I. R. Siti, "Study on demulsifier formulation for treating malaysian crude oil emulsion," Tech. Rep., Department of Chemical Engineering, Universiti Technologi Malaysia, 2006.

3.   R. Espinoza and W. Kleinitz, "The impact of Hidden Emulsion on Oil Prooducing wells—stimulation concept and field result," in Proceedings of the SPE European Formation Damage, The Hague, The Netherlands, 2003, SPE paper 00082252.

4.   F. Merv and F. Ben, "Studies of the formation process of water-in-oil emulsions," Marine Pollution Bulletin, vol. 47, no. 9–12, pp. 369–396, 2003.

5.   K. P. Micheal, C. Shaokum, and C. M. Samuel, "The key to Predicting Emulsion stability: solid content," in Proceedings of the SPE International Symposium on Oil Field Chemistry, Houston, Tex, USA, 2005, SPE paper 93008.

6.   D. Christophe, A. David, S. Anne, G. Alain, and B. Patrick, "Stability of water/crude oil emulsions based on interfacial dilatational rheology," Journal of Colloid and Interface Science, vol. 297, no. 2, pp. 785–791, 2006.

7.   M. Rondón, J. C. Pereira, P. Bouriat, A. Graciaa, J. Lachaise, and J. L. Salager, "Breaking of water-in-crude-oil emulsions. 2. Influence of asphaltene concentration and diluent nature on demulsifier action,"Energy and Fuels, vol. 22, no. 2, pp. 702–707, 2008.

8.   E. J. Ekott and E. J. Akpabio, "Influence of asphaltene content on demulsifiers performance in crude oil emulsions," Journal of Engineering and Applied Sciences, vol. 6, no. 3, pp. 200–204, 2011.

9.   K. Sunil, A. Abdullah, and N. S. Meeranpillal, "An Investigative study of potential emulsion problems before field development," in Proceedings of the SPE Annual Technical Conference and Exhibition, San Antonio, Tex, USA, 2007, SPE paper 102856.

10. V. Efeovbokhan, T. Akinola, and F. Hymore, "Performance evaluation of formulated and commercially available de-emulsifiers," in Nigerian Society of Chemical Engineers Proceedings (NSChE ‹10), vol. 40, pp. 87–99, 2010.

11. D. T. Nguyen and N. Sadeghi, "Selection of the right demulsifier for chemical enhanced oil recovery," inInternational Symposium on Oilfield Chemistry, The Woodlands, Tex, USA, April 2011.

12. C. I. Oseghale, E. J. Akpabio, and G. Udottong, "Breaking of oil-water emulsion for the improvement of oil recovery operations in the Niger Delta Oilfields," International Journal of Engineering and Technology, vol. 2, no. 11, pp. 1–7, 2012.

13. B. Fu, "Flow assurance—a technological review of Managing fluid behaviour and solid deposition to Ensure optimum flow," in Proceedings of the 7th Annual International Forum for deepwater Technologies (Deeptec ‹00), Aberdeen, UK, January 2000.

14. C. Noïk, H. Malot, C. Dalmazzone, and A. Mouret, "Encapsulation of crude oil emulsions," Oil and Gas Science and Technology, vol. 59, no. 5, pp. 535–546, 2004.

15. M. Nuraini, H. N. Abdurahman, and A. M. S. Kholijah, "Effect of chemical breaking agents on water-in crude oil emulsion system," International Journal of Chemical and Environmental Engineering, vol. 2, no. 4, pp. 1–5, 2011.

16. K. P. Micheal, C. Shaokum, A. M. Robert, and C. M. Samuel, "Classifying crude oil emulsion using chemical demulsifiers and stastical analyses," in Proceedings of the SPE Annual Technical Conference and Exhibition, Denver, Colo, USA, 2003, SPE paper 84610.

17. S. D. Taylor, "Investigations into the Electrical and rheological Behaviour of W/O—emulsions in high voltage Gradients," Colloid & Surfaces, vol. 29, pp. 25–51, 1988.

18. D. G. Thompson, A. S. Taylor, and D. E. Graham, "Emulsification and demulsification related to crude oil production," Colloid & Surfaces, vol. 15, pp. 175–189, 1987.

19. T. J. Jones, E. L. Neustadter, and K. P. Whittingham, "Water-in-crude oil emulsion stability and emulsion destabilization by chemical demulsifiers," Journal of Canadian Petroleum Technology, vol. 17, no. 2, pp. 100–108, 1978.

20. N. H. Abdurahman and W. K. Mahmood, "Stability of water-in-crude oil emulsions: effect of cocamide diethanolamine (DEA) and Span 83," International Journal of Physical Sciences, vol. 7, no. 41, pp. 5585–5597, 2012.

21. J. D. McLean and P. K. Kilpatrick, "Effects of asphaltene solvency on stability of water-in-crude-oil emulsions," Journal of Colloid and Interface Science, vol. 189, no. 2, pp. 242–253, 1997. ·

22. J. H. Norman, Non-Technical Guide to Petroleum, Geology, Exploration, Drilling and Production, Penswell Corporation, Tulsa, Okla, USA, 2nd edition, 2001.

Chapter 3

An Improved Negative Pressure Wave Method for Natural Gas Pipeline Leak Location Using FBG Based Strain Sensor and Wavelet Transform

Qingmin Hou[1], Liang Ren[2], Wenling Jiao[1], Pinghua Zou[1], and Gangbing Song[2,3]

[1]School of Municipal and Environmental Engineering, Harbin Institute of Technology, Harbin 150090, China

[2]Faculty of Infrastructure Engineering, Dalian University of Technology, Dalian 116024, China

[3]Department of Mechanical Engineering, University of Houston, Houston, TX 77004, USA

ABSTRACT

Methods that more quickly locate leakages in natural gas pipelines are urgently required. In this paper, an improved negative pressure wave method based on FBG based strain sensors and wavelet analysis is proposed. This method takes into account the variation in the negative pressure wave propagation velocity and the gas velocity variation, uses the traditional leak location formula, and employs Compound Simpson and Dichotomy Searching for solving this formula. In addition, a FBG based strain sensor instead of a traditional pressure sensor was developed for detecting the negative pressure wave signal produced by leakage. Unlike traditional sensors, FBG sensors can be installed anywhere along the pipeline, thus leading to high positioning accuracy through more frequent installment of the sensors. Finally, a wavelet transform method was employed to locate the pressure drop points within the FBG signals. Experiment results show good positioning accuracy for natural gas pipeline leakage, using this new method.

INTRODUCTION

Nowadays, pipelines have become ubiquitous for natural gas transportation. Therefore, natural gas pipelines play a vital role in modern enterprises and economies. However, leakage from natural gas pipelines occurs due to inevitable factors such as pipeline aging, erosion, natural disaster, and third party intrusion. Without proper and immediate fixing, leakage can lead to serious pollution and the danger of explosion due to the poisonous and explosive properties of natural gas. Therefore, effective ways to detect the exact position of leakage accidents are important, so that losses and danger can be greatly reduced.

Over recent years, the number of techniques for leak location of natural gas pipeline has been grown [1]. Presently, methods based on control theory and signal processing are popular, such as the methods based on model detection, pressure gradient, and methods based on negative pressure waves [2, 3]. With the advantage of quick response speeds, the negative pressure wave method based on pressure sensors is the most widely used leak location technology [4]. However, the installation of pressure sensors required for this method necessitates localized deconstruction of the pipeline. So pressure sensors are usually installed at input and output points of a pipeline and seldom anywhere in between. This leads to large signal attenuation and interference, leading to high rates of false alarms whilst reducing the precision of locating algorithms. In addition, the traditional leak location formula often assumes that the propagation velocity of the negative pressure wave and the velocity of natural gas in the pipeline are constants, even ignoring the velocity of natural gas. This does not match with the actual situation and will inevitably result in a large positioning error. Therefore, research is required into developing an improved method to overcome these shortcomings.

Distributed fiber optical sensors have been widely used for leak detection and location of natural gas pipeline [5–7]. The simple principle of this method is that a leak in a natural gas pipeline will lead to temperature variation or vibration, which can be detected by a distributed fiber optical sensor. The position of leakage point can be obtained by processing the captured signal. Temperature variation or vibration can be caused by other factors such as environment temperature change or random vibration of fiber optical. Thus vulnerable to interference and high false positive rate are main drawbacks of this method.

Fiber Bragg grating (FBG) sensor offers a number of advantages over traditional sensor, including immunity to electromagnetic interference, being light weight and durable, having small size, incorporating multiplexing capabilities, and is easy to install [8–14]. Due to the above attractive application features, FBG sensors have been playing an increasingly important role in the sensing community and have been widely used in structural health monitoring, damage detection, aviation, and other fields [15–21]. However, FBGs have not been extensively adopted in natural gas pipeline leakage location. While a kind of swellable polymer based FBG strain sensor has been proposed for oil pipeline leak detection and location [22]. The main component of natural gas is methane, the chemical properties of methane are very stable, and only a scarce number of polymer types swell when encountering methane. Therefore, this method is not suitable for natural gas pipeline. It is necessary to develop a FBG sensor applicable to natural gas pipelines.

In this paper, firstly, a modified leakage location formula was proposed based on the principle of negative pressure wave location, and Compound Simpson formula and Dichotomy Searching were employed to solve this formula. Secondly, a FBG based strain sensor for collecting the negative pressure wave signals was developed and experimentally tested. Finally, in order to get accurate time difference for that formula, a wavelet transform method was demonstrated to identify the pressure drop point within the FBG signals.

MODIFIED LEAK LOCATION FORMU-LA

Principle of Negative Pressure Wave Leak Location and Traditional Leak Location Formula

When leaks develop in a natural gas pipeline, the gas density near the leaking point will decrease rapidly. This phenomenon results in a negative pressure wave, which propagates through the pipeline from the leak point. Pressure sensors installed upstream and downstream can collect such negative pressure wave signals. According to time difference for detected signals and propagation velocity in the medium, the exact position of the leakage can be calculated [23]. The principle for negative pressure wave propagation is now described, in relation to Figure 1.

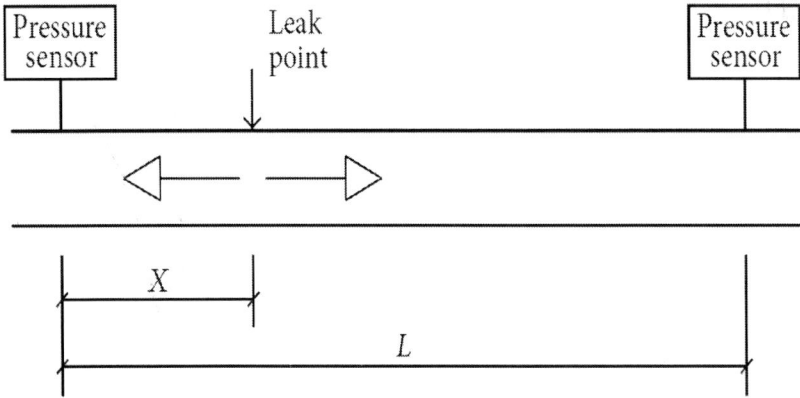

Figure 1: Schematic of negative pressure wave propagation.

In regard to Figure 1, assume that the distance between two sensors is L, the propagation velocity of negative pressure wave in the pipeline is v, the distance between the leak point and upstream sensor is x, the times when the wave is detected by the two sensors are t_1, t_2, and the velocity of natural gas in the pipeline is u.

As the pipeline diameter and gas transportation velocity increase, the velocity of natural gas cannot be ignored compared with that of the negative pressure wave. With the velocity of natural gas taken into consideration in our study, the relations between the length and time variables can be developed as follows:

$$t_1 = \frac{x}{(v-u)},$$

$$t_2 = \frac{(L-x)}{(v+u)},$$

$$\Delta t = t_1 - t_2.$$

(1)

The distance between the leak point and upstream sensor can be obtained from (1):

$$x = \frac{1}{2v}\left[L(v-u) + \Delta t \left(v^2 - u^2\right) \right].$$

(2)

Equation (2) is the traditional leak location formula.

Modification of Leak Location Formula

The traditional formula assumes that the propagation velocity of the negative pressure wave v, and the velocity of natural gas in the pipeline, u, are constants. In fact, v and v are related with the temperature, pressure, density, and specific heat of the surrounding medium, and the formulas of these two velocities can be obtained by thermal and hydraulics analysis. Treating the propagation velocity of the negative pressure wave and natural gas as variable parameters, the leak location formula can be rewritten as

$$t_1 = \int_0^x \frac{1}{v(x) - u(x)}\, dx,$$

(3)

$$t_2 = \int_x^L \frac{1}{v(x) + u(x)}\, dx,$$

(4)

$$\Delta t = t_1 - t_2.$$

(5)

Since the expressions for V(x) and u(x) are complex, the integral above is not a simple definite integral. Therefore, numerical integral was chosen for determining t_1, t_2. The Compound Simpson formula was employed to calculate the variable integral as given above. As can be seen from the above equations, if a value for the leak position, x, is assumed, the propagation velocity of negative pressure wave V(x) and the velocity of natural gas in the pipeline u(x) can be calculated. Finally, the time difference Δt can be determined, from which the leak position x can be improved. This is a recursive formula, as x cannot be solved deterministically. Therefore, a Dichotomy Searching was employed to solve this problem.

The Methods of Solving the Modified Formula

Compound Simpson Formula

The integral domain [a,b] is divided into n equal parts. The approximate integral value of the function f(x) in this domain can be obtained by

$$\int_a^b f(x)\, dx = \frac{h}{6} \sum_{i=0}^{n-1} \left[f(x_i) + 4\left(x_{i+1/2}\right) + f\left(x_{i+1}\right) \right],$$

(6)

where $x_i = a + ih$ ($i = 0, 1, \ldots, n$) and $h = (b - a)/n$, which is the step length. Equation (6) is the Compound Simpson formula. Using this formula to solve (3) and (4), the time difference t can be obtained as

$$\Delta t = t_1 - t_2$$

$$= \int_0^x \frac{1}{v(x) - u(x)}\, dx - \int_x^L \frac{1}{v(x) - u(x)}\, dx$$

$$\approx \frac{h}{6} \sum_{i=x}^{x-1} \left[\frac{1}{v(x_i) - u(x_i)} + 4\left(\frac{1}{v(x_{i+1/2}) - u(x_{i+1/2})} \right) \right.$$

$$\left. + \frac{1}{v(x_{i+1}) - u(x_{i+1})} \right]$$

$$- \frac{h}{6} \sum_{i=x}^{L-1} \left[\frac{1}{v(x_i) + u(x_i)} + 4\left(\frac{1}{v(x_{i+1/2}) + u(x_{i+1/2})} \right) \right.$$

$$\left. + \frac{1}{v(x_{i+1}) + u(x_{i+1})} \right].$$

(7)

Dichotomy Searching

In order to locate the leaking point, Dichotomy Searching was used to determine x. Dichotomy Searching is explained below in reference to Figure 2.

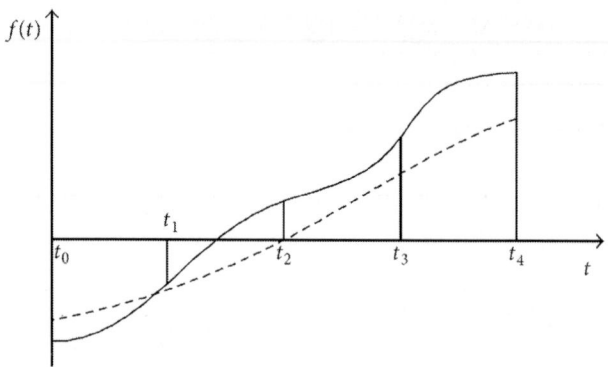

Figure 2: Sketch map of the Dichotomy Searching.

As shown in Figure 2, the midpoint $t2$ of domain $[t_0, t_4]$ is obtained to calculate $f(t_2)$, then $f(t_0)$, $f(t_2)$, and $f(t_4)$ are compared (in case of $f(t_0)<0, f(t_4>0))$.

If $(t_2)>0$, then $[t_2, t_4]$ is rejected, and Dichotomy Searching is going in the domain of $[t_0, t_2]$.

The next computational point is the midpoint of domain $[t_0, t_2]$. (2) If $f(t_2)$

$$f(x) = \frac{h}{6}\sum_{i=0}^{x-1}\left[\frac{1}{v(x_i) - u(x_i)} + 4\left(\frac{1}{v(x_{i+1/2}) - u(x_{i+1/2})}\right)\right.$$

$$\left. + \frac{1}{v(x_{i+1}) - u(x_{i+1})}\right]$$

$$-\frac{h}{6}\sum_{i=x}^{L-1}\left[\frac{1}{v(x_i) + u(x_i)}\right.$$

$$+ 4\left(\frac{1}{v(x_{i+1/2}) + u(x_{i+1/2})}\right)$$

$$\left. + \frac{1}{v(x_{i+1}) + u(x_{i+1})}\right] - \Delta t.$$

$$(8)$$

The root of equation $(x) = 0$ is just the leak position, and the flow chart of solving process is shown in Figure 3.

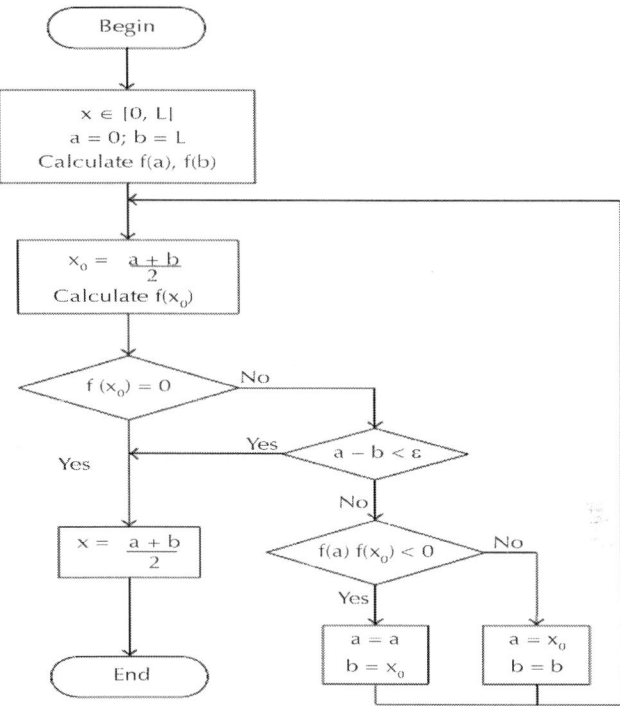

Figure 3: Flow chart of solving process.

Critical Factors of Follow-Up Work

From (8), it can be clearly observed that the actual detection time difference plays an important role in leak location. Two critical factors for precise measurements of the time difference have been identified as follows:

- sensors that can be easily installed, such that the spacing between sensors can be kept small,

- accurate identification of the pressure drop point (t_1, t_2) from the sensor pressure trace, as this directly influences the sensitivity and reliability of leakage locating.

The following two sections address these critical factors.

FBG BASED STRAIN SENSOR AND EXPERIMENT

Principle of FBG Based Strain Sensor

FBG based strain sensors are wrapped around the wall of a pipeline, as shown in Figure 4. A change in pressure within the pipeline leads to its expansion or contraction with the hoop (circumferential) strain of the pipeline changing accordingly. The FBG strain sensors detect pressure changes within the pipe by sensing the hoop strain. The hoop strain within a pipeline system can be expressed as

$$\varepsilon_y = \frac{\sigma_y - \upsilon\sigma_z}{E},$$

(9)

where ε_y is the pipeline hoop strain, υ is the pipeline Poisson ratio, σ_y is the pipeline hoop stress, σ_z is the pipeline axial stress, and E is the pipeline elasticity modulus.

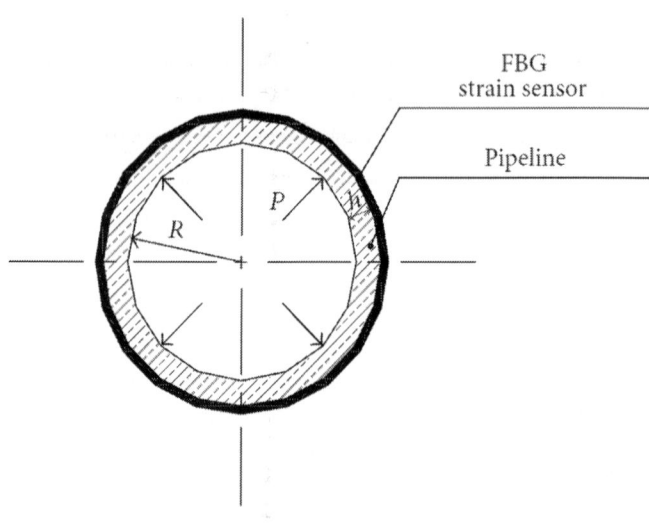

(a)

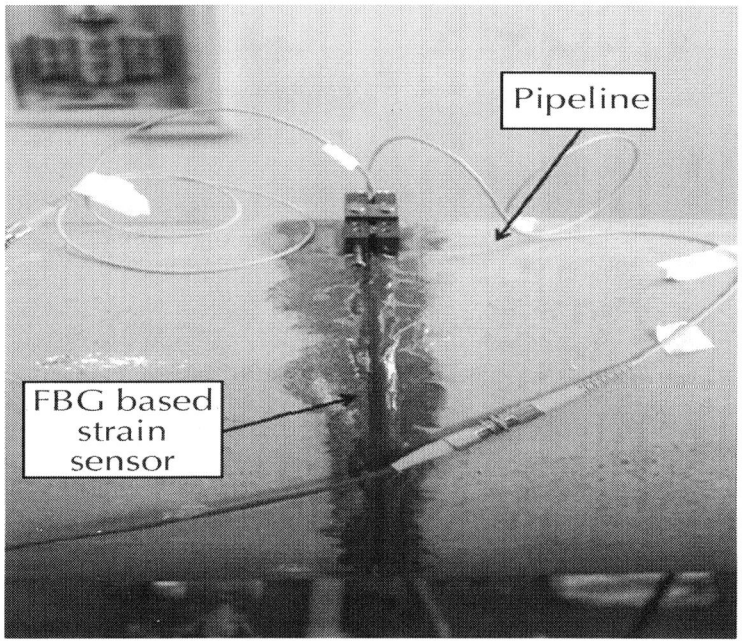

(b)

Figure 4: (a) Schematic of a FBG based strain sensor as installed on a pipeline; (b) photo of the sensor installed on an experimental pipeline.

Using (9), a relationship can be derived to relate the hoop strain with the pipeline pressure and pipe wall thickness. First, it is assumed that the pipeline is infinitely long, so that axial stress can be neglected; that, is $\sigma_z = 0$. Meanwhile, as $\sigma_y = pR/h$, the values for σ_y and σ_z can be substituted into (9), which gives

$$\varepsilon_y = \frac{pR}{hE},$$

(10)

where P is the pressure in the pipeline, R is the pipeline internal radius, and h is the pipeline wall thickness. As seen from (10), as the pressure in the pipeline changes, the pipeline hoop strain also changes linearly. Therefore FBG based strain sensors can detect the pressure variation by monitoring the hoop strain of the pipe wall.

Experimental Setup

The proposed methodology for detecting and locating gas pipeline leakage was tested on an experimental pipeline. The schematic for this gas pipeline is shown in Figure 5(a). Two air tanks and a section of pipeline were used to simulate a realistic gas transfer main. The pipeline in this experiment is made of steel, with a diameter of 273?mm as frequently used in practice. The pipeline length was 11?m due to lab space limitations. A leak point was simulated by manually opening valve at locations. As shown in Figure 5(b), a rotameter was located at the leak point for measuring the leak rate, and two FBG strain sensors (L1, L2) were installed.

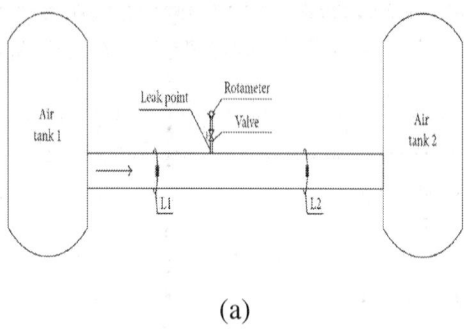

(a)

(b)

Figure 5: (a) Schematic diagram of experiment setup; (b) photo of experiment setup.

Briefly, the experimental process followed these steps: First, air is compressed into air tank 1 by air compressor. Air tank 1 plays the role of stabilizing the pressure in the pipeline. Second, when the pressure is stable, a valve is opened to simulate a leak in the pipeline. In the meantime, all the sensors gather data.

Experimental Results

The signals from FBG based strain sensors were captured to determine their ability to sense negative pressure waves caused by sudden leaks. A leakage was simulated by opening the valve at the leak point as shown in Figure 6(a). Sensors set upstream and downstream collected the hoop (circumferential) strain response, and the pipeline pressure was calculated from this signal. As seen in Figure 6, the pressure was steady before the leak occurred. A sudden pressure drop developed in the waveform that resulted from the leak. Because the duration of the leak was limited, so the pipeline finally returned to a steady but lower pressure due to the loss of gas. Furthermore, the waveforms recorded at L1 and L2 are similar, as shown in Figures 6(a) and 6(b); this was expected due to the propagation of the negative pressure wave on either side of the leak. These results indicate that the FBG based strain sensors can accurately detect the negative pressure wave produced by leakage events.

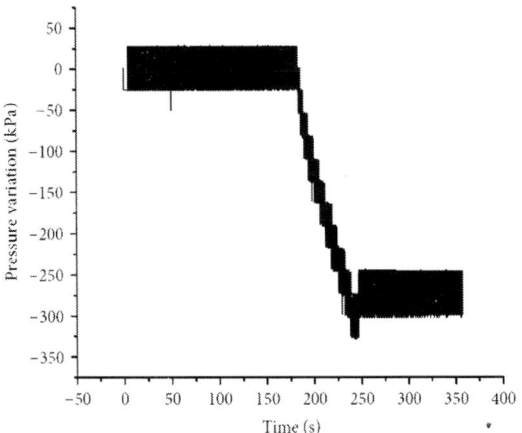

(a)

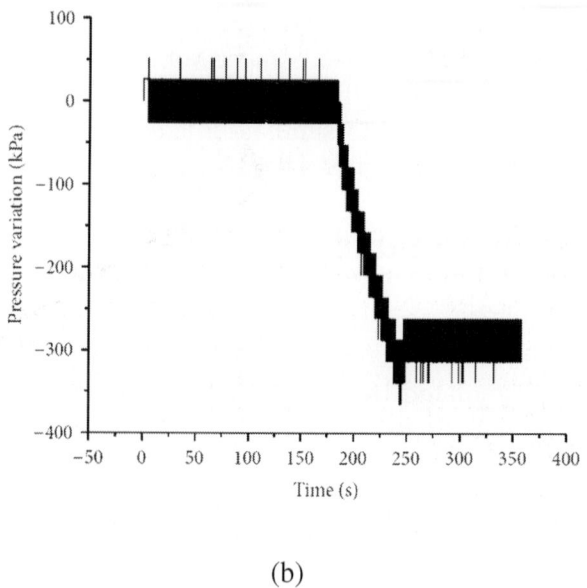

(b)

Figure 6: (a) Pressure measured by FBG based strain sensor L1; (b) pressure measured by FBG based strain sensor L2.

IDENTIFY THE PRESSURE DROP POINTS USING WAVELET TRANSFORM

Definition of Wavelet Transform

The wavelet transform method was used to find the sharp transition in the FGB trace, corresponding to the movement of the negative pressure wave passes through the FBG sensor location. This method was used because this method can scale up the signal to find the sharp transition simply.

The definition of wavelet transform for a function x(t) is given in [24–26]

$$WT_x(a,b) = \int x(t)\, \psi^* \frac{[(t-b)/a]}{\sqrt{a}}\, dt$$

$$= \int x(t)\, \psi^*_{a,b}(t)\, dt = \langle x(t), \psi_{a,b}(t) \rangle.$$

(11)

In this equation, a, b, and t are continuous variables, which is also the reason why (11) is called the continuous wavelet transform. If the wavelet coefficients are computed on all possible scales, the computational burden is large. With the consideration of practical feasibility in numerical computation and simplicity of theoretical analysis, wavelet transforms are normally discretized in practice. The relevant discrete wavelet transform is described by

$$C_{j,k} = \int_{-\infty}^{+\infty} f(t)\, \psi^*_{j,k}(t)\, dt = \langle f, \psi_{j,k} \rangle,$$

(12)

where $\psi_{j,k}(t)$ can be expressed in the following form:

$$\psi_{j,k}(t) = \frac{a_0 \psi\left(t - k a_0^j b_0\right)}{a_0^j}$$

$$= a_0^{-1/2} \psi\left(a_0^{-j} t - k b_0\right).$$

(13)

In practice, wavelet in (13) is usually dyadic, which means that

$$\psi_{j,k} = 2^{-j/2} \psi\left(2^{-j} t - k\right) j, \quad k \in z.$$

(14)

The Application of Wavelet Transform and Positioning Results

The wavelet transform maxima in modulus on all possible transform scales correspond to the positions where the signals have sharp transition [27]. So the singularity can be obtained by detecting the wavelet transform maxima. Further, the wavelet transform modulus of real singularity is almost fixed value on all scales. In contrast, the modulus of faked singularities is inversely proportional to the scale. Consequently, it is reasonable to determine the real singularity by using this property. However, the detection of singularity is more accurate on a small scale, but this process is likely to be interrupted by noise, As a result, the faked singularity may appear. On the other hand, in the large scale, although noise has little influence on the detection, the deviation between the real singularity and the detective one is large. Since practical pressure signals are mixed with noise even after filtering, it is beneficial to consider multiscale transforms to find the real singularity. Therefore, this study used the following procedure to determine the location of singularities. Firstly, the approximate range wherein the singularity was located was determined by large scales. Secondly, the real singularity was located within this range by taking advantage of the small scales.

In this experiment, the negative pressure wave was produced by opening a valve to simulate a leakage, and the magnitude of the pressure wave was measured by FBG based strain sensors over time. The FBG sensors trace at L1 and L2 were chosen to calculate the leakage position, and the distance between them is 8?m. The denoised signals at L1 and L2 are shown in Figure 7. The pressure drop points at both L1 and L2 are marked in Figure 7, and these points were used to determine the time difference using the methodology as describe above.

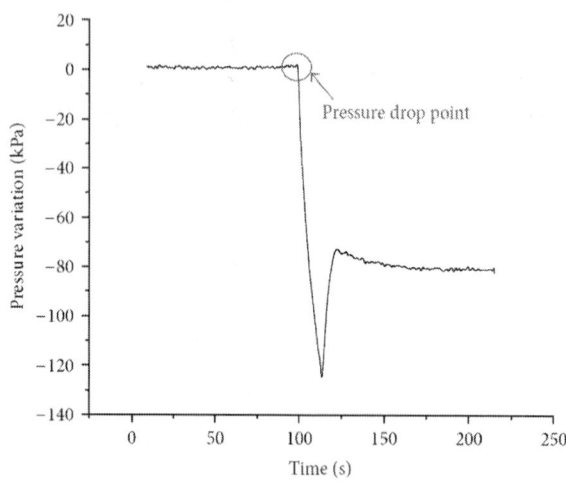

(a)

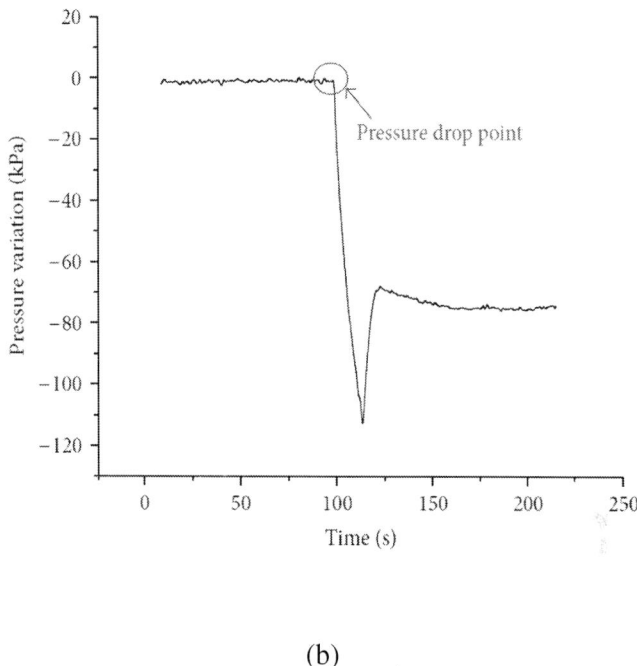

(b)

Figure 7: (a) Denoised pressure signal measured by FBG based strain sensor L1; (b) denoised pressure signal measured by FBG based strain sensor L2.

Figure 8 shows the wavelet analysis of the signal measured at L1 and L2. The pressure drop points in Figure 8are the singularities. In Figure 8(a) the detail signals at L1 after wavelet transform from scales of 9 to 12 are shown. Similarly, the results from applying wavelet transforms to the pressure signals from L2 are shown in Figure 8(b). When scale = 9, it can be clearly observed that there are many fake singularities in the detail signal, and this phenomenon also happens when the scale is less than 8. On the other hand, as the scale increases, the fake singularities are become less frequent, and the real singularities become more conspicuous, although singularity deviation may happen. Taking advantage of the multiscale transform, the singularities can be identified; therefore, the difference between the arrival time of the negative pressure wave between L1 and L2 can be determined. For this case study, this time difference was calculated to be $t \approx -1.24 \times 10^{-4}$ s. By substituting this value of t into (8), the distance between the leak point and L1 was calculated as x=3.98?m, whilst the actual distance was 3.6?m, yielding an absolute error of 0.38?m and a relative error of 4.8%. Because the distance between L1 and L2 in this experiment is very short (8?m), t₁ and t₂ are very similar, so small deviations in determining t cause large errors in relative positioning accuracy. However, this limitation would not apply to systems where the distance between L1 and L2 were greater as would be the case in practice. Therefore, the authors have reason to believe that this method can locate the leakage with good accuracy.

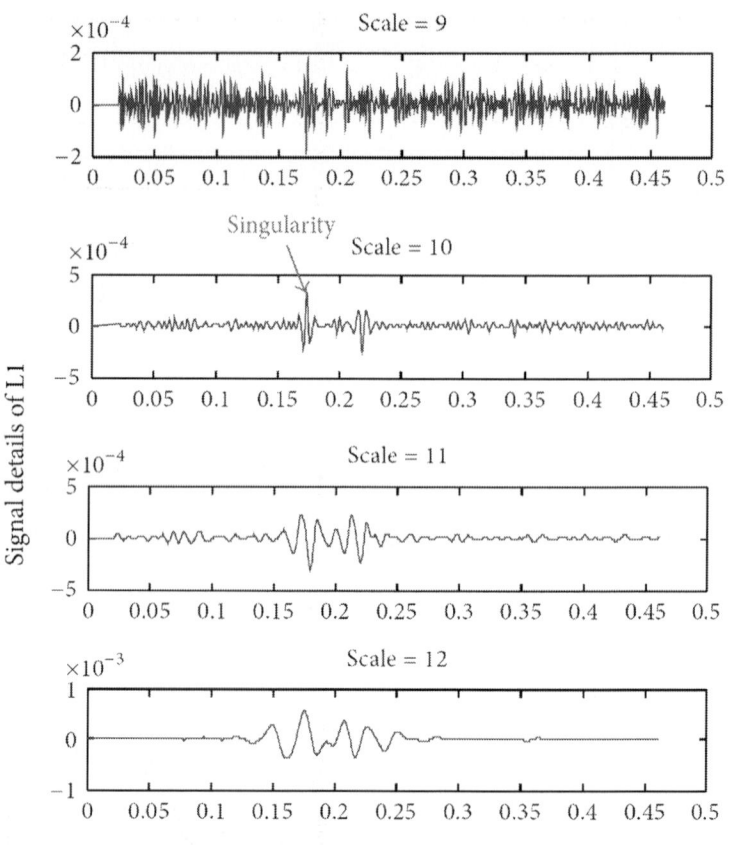

(a)

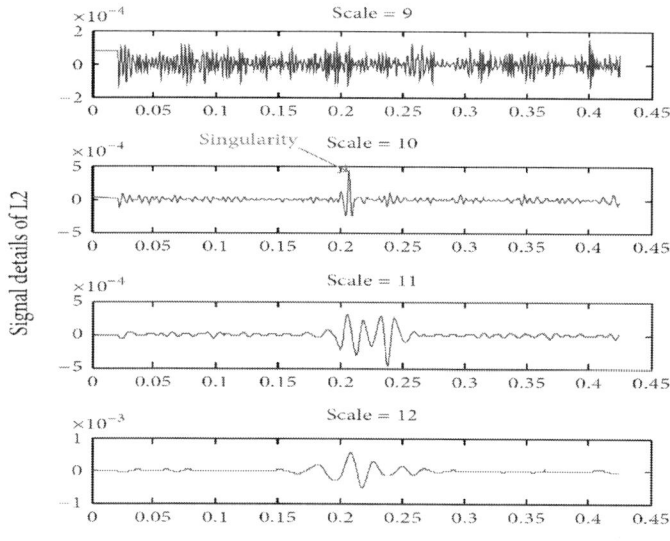

(b)

Figure 8: (a) Wavelet analysis of denoised signal measured by L1; (b) wavelet analysis of denoised signal measured by L2.

CONCLUSIONS

At present, the negative pressure wave method based on pressure sensors is the most widely used leak location technology. In this study, this method is enhanced through incorporating the variation of negative pressure wave and natural gas velocities into the negative pressure wave leak location formula. The Compound Simpson formula and Dichotomy Searching were employed to solve this modified formula. In order to overcome the installation difficulty of traditional pressure sensors, a FBG based strain sensor for collecting the negative pressure wave signals was developed and experimentally tested. Compared to conventional pressure sensors, FBG based strain sensors have favorable properties, such as high sensitivity, cheap cost, and ease of installation. Furthermore, a wavelet transform based method for identifying the pressure drop points within the FBG signals was proposed to calculate the leak position. Utilizing the above methods to calculate the leak position, an absolute error of 0.38?m was obtained in this experiment. This result demonstrated good positioning accuracy using this improved method.

ACKNOWLEDGMENTS

This work was partially supported by the Science Fund for Creative Research Groups from the National Science Foundation of China under Grant no. 51121005.

REFERENCES

1.   R. A. Silva, C. M. Buiatti, S. L. Cruz, and J. A. F. R. Pereira, "Pressure wave behaviour and leak detection in pipelines," Computers and Chemical Engineering, vol. 20, no. 1, pp. S491–S496, 1996.

2.   T.-H. Yi, H.-N. Li, and M. Gu, "Characterization and extraction of global positioning system multipath signals using an improved particle-filtering algorithm," Measurement Science and Technology, vol. 22, no. 7, Article ID 075101, 2011.

3.   T.-H. Yi, H.-N. Li, and X.-Y. Zhao, "Noise smoothing for structural vibration test signals using an improved wavelet thresholding technique," Sensors, vol. 12, no. 8, pp. 11205–11220, 2012.

4.   I. R. Ellul, "Advances in pipeline leak detection techniques," Pipes and Pipelines International, vol. 34, no. 3, pp. 7–12, 1989.

5.   B. Vogel, C. Cassens, A. Graupner, and A. Trostel, "Leakage detection systems by using distributed fiber optical temperature measurement," in Smart Structures and Materials 2001: Sensory Phenomena and Measurement Instrumentation for Smart Structures and Materials, Proceedings of SPIE, pp. 23–34, Newport Beach, Calif, USA, March 2001.

6.   E. A. Mendoza, R. A. Lieberman, J. Prohaska, and D. Robinson, "Distributed fiber optic chemical sensors for detection of corrosion in pipelines and structural components," in Nondestructive

Evaluation of Utilities and Pipelines II, Proceedings of SPIE, pp. 136–143, San Antonio, Tex, USA, April 1998.

7.   D. S. McKeehan, R. W. Griffiths, and J. E. Halkyard, "Marine applications for a continuous fiber optic strain monitoring system," in Proceedings of the 18th Annual Offshore Technology Conference, pp. 342–345, Houston, Tex, USA, 1986.

8.   K. T. V. Grattan and B. T. Meggitt, Optical Fiber Sensor Technology, vol. 2, Chapman and Hall, London, UK, 1998.

9.   K. T. V. Grattan and B. T. Meggitt, Optical Fiber Sensor Technology, vol. 3-4, Kluwer Academic, Boston, Mass, USA, 1999.

10. B. Culshaw and J. Dakin, Optical Fiber Sensors, vol. 1–4, Artech House, Boston, Mass, USA, 1988–1997.

11. A. Dandridge and C. Kirkendall, "Passive fiber optic sensor networks," in Handbook of Optical Fiber Sensing Technology, J. M. Lopez-Higuera, Ed., pp. 433–448, Wiley, New York, NY, USA, 2002.

12. A. Kersey, "Distributed and multiplexed fiber optic sensors," in Fiber Optic Sensors: An Introduction for Engineers and Scientists, E. Udd, Ed., pp. 325–368, Wiley, New York, NY, USA, 1991.

13. S. Yin, P. B. Ruffin, and F. T. S. Yu, Fiber Optic Sensors, CRC Press, Boca Raton, Fla, USA, 2nd edition, 2008.

14. O. S. Wolfbeis, "Fiber-optic chemical sensors and biosensors," Analytical Chemistry, vol. 78, no. 12, pp. 3859–3873, 2006.

15. S. Yashiro, T. Okabe, and N. Takeda, "Damage identification in a holed CFRP laminate using a chirped fiber Bragg grating sensor," Composites Science and Technology, vol. 67, no. 2, pp. 286–295, 2007.

16. T.-H. Yi and H.-N. Li, "Methodology developments in sensor placement for health monitoring of civil infrastructures," International Journal of Distributed Sensor Networks, vol. 2012, Article ID 612726, 11 pages, 2012.

17. H. Tsuda, J.-R. Lee, and Y. Guan, "Fatigue crack propagation monitoring of stainless steel using fiber Bragg grating ultrasound sensors," Smart Materials and Structures, vol. 15, no. 5, article no. 032, pp. 1429–1437, 2006.

18. D. C. Betz, G. Thursby, B. Culshaw, and W. J. Staszewski, "Advanced layout of a fiber Bragg grating strain gauge rosette,"

Journal of Lightwave Technology, vol. 24, no. 2, pp. 1019–1026, 2006.

19. H.-J. Park and M. Song, "Linear FBG temperature sensor interrogation with Fabry-Perot ITU multi-wavelength reference," Sensors, vol. 8, no. 10, pp. 6769–6776, 2008.

20. N. Takeda, Y. Okabe, J. Kuwahara, S. Kojima, and T. Ogisu, "Development of smart composite structures with small-diameter fiber Bragg grating sensors for damage detection: quantitative evaluation of delamination length in CFRP laminates using Lamb wave sensing," Composites Science and Technology, vol. 65, no. 15-16, pp. 2575–2587, 2005.

21. T.-H. Yi, H.-N. Li, and H.-M. Sun, "Multi-stage structural damage diagnosis method based on "energy-damage" theory," Smart Structures and System, vol. 12, no. 3-4, pp. 345–361, 2013.

22. R. M. López, V. V. Spirin, M. G. Shlyagin et al., "Coherent optical frequency domain reflectometry for interrogation of bend-based fiber optic hydrocarbon sensors," Optical Fiber Technology, vol. 10, no. 1, pp. 79–90, 2004.

23. X. Lu, Y. Sang, J. Zhang, and Y. Fan, "A pipeline leakage detection technology based on wavelet transform theory," in Proceedings of IEEE International Conference on Information Acquisition (ICIA '06), pp. 1432–1437, Shandong, China, August 2006.

24. I. Daubechies, "The wavelet transform, time-frequency localization and signal analysis," IEEE Transactions on Information Theory, vol. 36, no. 5, pp. 961–1005, 1990.

25. H. Li, T. Yi, M. Gu, and L. Huo, "Evaluation of earthquake-induced structural damages by wavelet transform," Progress in Natural Science, vol. 19, no. 4, pp. 461–470, 2009.

26. T.-H. Yi, H.-N. Li, and M. Gu, "Wavelet based multi-step filtering method for bridge health monitoring using GPS and accelerometer," Smart Structures and Systems, vol. 11, no. 4, pp. 331–348, 2013.

27. Y. Hao, W. Guizeng, and F. Chongzhi, "Application of wavelet transform to leak detection and location in transport pipelines," Engineering Simulation, vol. 13, no. 6, pp. 1025–1032, 1996.

Advances in Asset Management Techniques: An Overview of Corrosion Mechanisms and Mitigation Strategies for Oil and Gas Pipelines

Chinedu I. Ossai

Production Planning Department, Overall Forge Pty Ltd, 70 R W Henry Drive, Ettamogah near Albury, Albury, NSW 2640, Australia

ABSTRACT

Effective management of assets in the oil and gas industry is vital in ensuring equipment availability, increased output, reduced maintenance cost, and minimal nonproductive time (NPT). Due to the high cost of assets used in oil and gas production, there is a need to enhance performance through good assets management techniques. This involves the minimization of NPT which accounts for about

20–30% of operation time needed from exploration to production. Corrosion contributes to about 25% of failures experienced in oil and gas production industry, while more than 50% of this failure is associated with sweet and sour corrosions in pipelines. This major risk in oil and gas production requires the understanding of the failure mechanism and procedures for assessment and control. For reduced pipeline failure and enhanced life cycle, corrosion experts should understand the mechanisms of corrosion, the risk assessment criteria, and mitigation strategies. This paper explores existing research in pipeline corrosion, in order to show the mechanisms, the risk assessment methodologies, and the framework for mitigation. The paper shows that corrosion in pipelines is combated at all stages of oil and gas production by incorporating field data information from previous fields into the new field's development process.

INTRODUCTION

The oil and gas industry is an asset intensive business with capital assets ranging from drilling rigs, offshore platforms and wells in the upstream segment, to pipeline, liquefied natural gas (LNG) terminals, and refineries in the midstream and downstream segments. These assets are complex and require enormous capital to acquire. An analysis of the five major oil and gas companies (BP, Shell, ConocoPhillips, Exxonmobil, and Total) shows that plant, property, and equipment on average accounts for 51% of the total assets with a value of over $100 billion [1]. Considering the huge investment in assets, oil and gas companies are always under immense pressure to properly manage them. To achieve this involves the use of different optimization strategies that is aimed at cost reduction and improved assets reliability [2].

Due to the growth in the demand of oil and gas around the world, companies are developing new techniques to reach new reservoirs in the offshore and onshore arena [3]. This is putting pressure on most of the facilities with the attendant cost of maintenance soaring [1]. The continuous utilization and the ageing of facilities have resulted in record failures in the oil and gas plants. Research shows that between 1980 and 2006, 50% of European, major hazards of loss containment events arising from technical plants failures were primarily due to ageing plants mechanism caused by corrosion, erosion, and fatigue [4, 5].

A study shows that corrosion cost in US rose above 1 $ trillion in 2012 accounting for about 6.2% of GDP hence, the largest single expense in the economy [6]. In the oil and gas company, corrosion accounts for over 25% of assets failure [7] and is found to be prevalent in every stage of the production cycle. Oxygen which plays a dominant role in corrosion is normally present in producing formation water. During drilling operation, drilling mud can corrode the well casing, drilling equipment, pipeline, and the environment. Water and CO_2 produced or injected for secondary recovery can cause severe corrosion of completion strings, while the acids used to reduce formation damage around the well or to remove scale can attack metals [8]. The formation water and injected water used for the oil recovery are a potential source of pipeline corrosion during transportation of the oil from the wells to the loading terminals. Mechanical static equipment like valves, tanks, vessels, separators, and so forth are susceptible to a different kind of corrosion however, pipelines are more prone to corrosion due to the presence of CO_2, H_2S, H_2O, bacteria, sand, and so forth in the fluid.

Owing to the increasing cost of pipeline corrosion management in the oil and gas industries [1], operators are becoming more concerned about corrosion management planning at all phases of production. Corrosion information from existing field data is being incorporated into design information for new oil and gas field [9,10] in a bid to develop appropriate corrosion management methodologies that will enhance the design life of the pipelines and optimize production. To reduce the risk of microbiologically influenced Corrosion (MIC) and other associated corrosions like stress corrosion cracking (SCC), hydrostatic testing of carbon steel pipes should be carried out in such a manner that enhances the future pipeline service conditions by using the right source of water, ensuring proper degree of filtration, ensuring limited exposure period to temperature and eliminating air packets [11]. Though bacteria in the biofilm are responsible for pitting of a pipeline in a MIC however, the impact of the flow velocity of the constituent fluid influences the mass transfer rate thereby affecting the biofilm formation, hence, inhibiting the activities of sulphate reducing bacteria, (SRB) present in the fluid [12]. This flow attribute has significant impact in MIC in oil and gas pipeline.

Considering the fact that the CO_2 and H_2S induced corrosion rate can reach up to 6 mm/yr and 300 mm/yr, respectively, [13] in oil and gas pipelines, sophistication in inspection and monitoring techniques is therefore necessary for quick mitigation. The increased trend in in-line inspection and online data acquisition has helped in quicker data acquisition, analysis, and decision making regarding corrosion in pipelines. The enhanced research knowledge of the behaviour of these corrodents (CO_2 and H_2S, acetic acid, etc.) at different operating conditions [14–17] has given rise to numerous mechanistic, statistical, and empirical models [18–23] which have contributed immensely in the inspection and monitoring, selection of inhibitors, and materials selection for pipelines design.

Since corrosion is a dominant factor contributing to failures and leaks in pipelines [24], to aid industry experts in managing the integrity of pipelines therefore involves a layout of the developments in the management strategies. This involves the recognition of the conditions contributing to the corrosion incident and identifying effective measures that can be taken to mitigate against them. To facilitate best practices in pipeline integrity management therefore, requires a framework that utilizes good policies and procedures in inspection, data collection, and interpretation for corrosion control.

OVERVIEW OF CORROSION

Corrosion is a naturally occurring phenomena commonly defined as the deterioration of a substance (usually metal) or its properties because of a reaction with its environment [25]. Corrosion of materials is inevitable due to the fundamental need of lowering of Gibbs energy [26]. Every material is trying to achieve a lower energy state hence the ability to corrode in order to get to a low energy oxide state. Though this is the case with all materials, the major focus of experts however, is to achieve an equilibrium position between the materials and the environment thereby controlling corrosion.

Modern corrosion science has its roots in electrochemistry and metallurgy. Whereas electrochemistry contributes to the understanding of materials via corrosion, metallurgy provides information about the behaviour of the material and their alloys hence provide a medium for combating the degradation on them. The type of corrosion mechanism and its rate of attack depend on the nature of the environment (air, soil, water, etc.) in which the corrosion takes place. Whereas some environmental condition can help to mitigate the rate of corrosion, others help to increase it hence, industrial wastes and products can either be corrosion inhibitor or catalyst. For instance, CO_2, H_2S, temperature, mass flow rate, pH, formation water, and so forth contribute in no small measure to the rate of corrosion in oil and gas pipeline [14, 16, 17, 27]. The existence of anodic cathodic sites on the surface of a piece of metal implies that the difference in electrical potential is found on the surface. This potential difference has the tendency of initiating corrosion. If an oil and gas pipeline passes through a zone of clay soil (where the oxygen concentration is low) to gravel (where the oxygen concentration is high), the part of the pipeline in contact with the clay becomes anodic and suffers damage. Though this problem is extensively addressed with the cathodic protection [26], concentration cell may also be formed where there are differences in metal ion concentration.

Although most metals are crystalline in form, they generally are not continuous single crystal but rather are collections of small grains of domains of localized order in which microcrystal forms as the liquid cools and solidifies. In the final states, the crystals have different orientation with respect to one another. The edge of the domain form grain boundaries which are an example of planar defects in metal. These defects are usually sites of chemical reactivity. The boundaries are also weaknesses, the places where stress corrosion cracking begins. The metallic surface exposed to an aqueous electrolyte usually possesses site for oxidation (anodic reaction) that produces electrons in the metal and reduction (cathodic reaction) that consumes the electrons produced by the anodic reaction [25, 26]. These sites make up a corrosion cell. The anodic reaction (Figure 1) involves the dissociation of metal to form either soluble ionic product or an insoluble compound of metal usually an oxide. For cathodic reaction (Figure 2), oxygen gas generated could be reduced or water is reduced to produce hydrogen gas. The simultaneous reaction of the anodic and cathodic reactions produces the electrochemical cell.

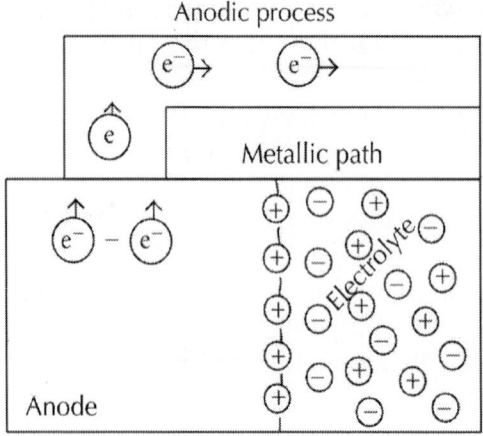

Figure 1: Anodic process.

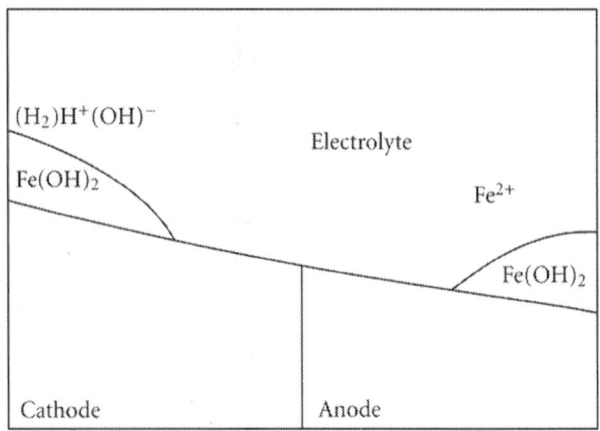

Figure 2: Cathodic process.

In completely oxygen-free water, the cathodic reaction that takes place is the reaction of hydrogen ion to form hydrogen gas as shown in (1):

$$2H^+ + 2e^- \longrightarrow H_2(g)$$

$$(1)$$

When significant amounts of oxygen are present in the system, the cathodic reaction that takes place is shown in (2):

$$2H^+ + \frac{1}{2}O_2 + 2e^- \longrightarrow H_2O$$
(2)

The hydrogen ion is present in water due to the ubiquitous dissolution of water into hydroxyl ions as shown in

$$2H_2O \longrightarrow 2H^+ + 2(OH)^-$$
(3)

(3):In the anode, there is a dissociation of iron to form a ferrous ion as shown in (4).

$$Fe \longrightarrow Fe^{2+} + 2e^-$$
(4)

The ferrous ion will react with the hydroxyl ion to form insoluble ferrous hydroxide as shown in (5):

$$Fe^{2+} + 2(OH)^- \longrightarrow Fe(OH)_2$$
(5)

The anodic and cathodic reactions that take place in a neutral and alkaline condition is shown in Figure 3.

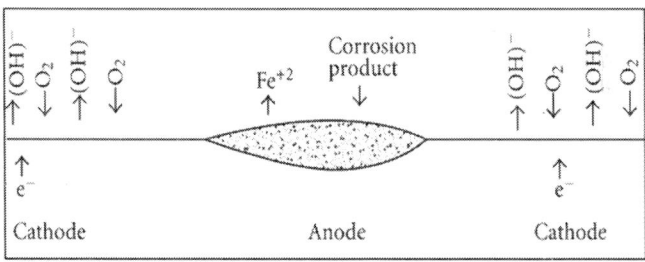

Figure 3: Neutral and Alkaline condition of a corrosion process.

The cathodic reaction is as follows:while the anodic reaction is the same as (4).

For an anodic condition, the cathodic and anodic reactions are represented in Figure 4.

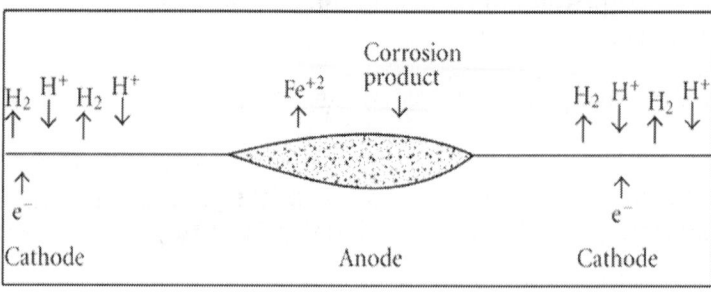

Figure 4: Acidic Condition of a Corrosion process.

The cathodic reaction equation is shown in (1),

$$\frac{1}{2}O_2 + H_2O + 2e^- \longrightarrow 2(OH)^-$$

(6)

while the anodic equation is shown in (4).

In a deoxygenated solution, the hydrogen reaction combines with the others to yield the net corrosion reaction shown in (7):

$$Fe + H_2O + \frac{1}{2}O_2 \longrightarrow Fe(OH)_2$$

(7)

In oxygenated aqueous systems, the oxygen reduction leads to a slightly different net corrosion reaction as shown in (8):

$$Fe + H_2O + \frac{1}{2}O_2 \longrightarrow Fe(OH)_2$$

(8)

Whereas in deoxygenated solution, hydrogen is evolved, in oxygenated system, oxygen is consumed. The evolved hydrogen acts as a catalyst for the formation of magnetite (Fe_3O_4) in deoxygenated water. Experiment shows that hydroxide readily decomposes into magnetite in deoxygenated water above 100°C [28] as indicated in (9):

$$3Fe(OH)_2 \longrightarrow Fe_3O_4 + H_2(g) + 2H_2O$$

(9)

The net corrosion reaction with the magnetite as the final product is shown in (10):

$$3Fe + 4H_2O \longrightarrow Fe_3O_4 + 4H_2(g)$$

(10)

In oxygenated solution, the ferrous oxide (Fe^{2+}) does not immediately precipitate out since it rapidly oxidizes to ferric oxide (Fe^{3+}), as a result, insoluble iron hydroxide is formed which is converted to hematite as shown in (11):

$$Fe(OH)_2 + \frac{1}{2}H_2O + \frac{1}{4}O_2 \longrightarrow Fe(OH)_3$$

(11)

The ferric oxide (Fe^{3+}) is converted to magnetite according to (12)

$$2Fe(OH)_3 + Fe(OH)_2 \longrightarrow Fe_3O_4 + 4H_2O$$

(12)

Cathodic and anodic sites could be built as a result of variation in environmental conditions, metallic microstructure variation, and variation in environmental concentration of oxygen at different points of a metal [26]. At the anodic sites, the dissolution of metallic ions in the electrolyte brings about the flow of electrons between the corroding anodes and non-corroding cathodes. The spontaneous nature of the corrosion however, depends on the rate of flow of these electrons.

Though establishing the tendency for corrosion is necessary, however, it is more important to determine the rate of corrosion. This is because a particular metal or alloy may be prone to corrosion in an environment but at a very low rate, in which it will not be a problem [26]. To understand the rate of corrosion however, requires the knowledge of the role of primary environment and metallurgical variables, underlying mechanism of corrosion, and synthesis of information to account for effects of the parameters.

MECHANISMS OF CORROSION IN OIL AND GAS PIPELINES

Fluid flowing from oil and gas pipelines has a combination of chemicals including CO_2, H_2S, organic acids, bacteria, sand, and water. These constituents are among the major causes of corrosion in pipeline. The CO_2 dissolves in the presence of water to form an acidic oxide which reacts with iron. This type of corrosion is referred to as sweet corrosion. This is responsible for most types of general corrosion in oil and gas pipeline. Sour corrosion occurs when H_2S in the excess of 100ppm is present in the oil and gas, causes corrosion in the pipeline, and predominantly causes pitting [26, 29].

CO_2 present in oil and gas will dissolve in water to produce carbonic acid (H_2CO_3) [23, 27]. This acid dissolves steel to produce iron carbonate and hydrogen as shown in (13). This reaction takes place at the cathode:

$$Fe + H_2CO_3 \longrightarrow FeCO_3 + H_2(g) \tag{13}$$

Despite the weakness of carbonic acid it is extremely corrosive to carbon steel. The chemical reactions above form the iron carbonate films. Depending on the condition during the formation, these films can be protective or non-protective at the anode, iron dissolves as shown in (4). The presence of CO_2 acts as a catalyst increasing the hydrogen evolution thereby increasing the corrosion rate of carbon steel in aqueous solution [27]. The carbonic acid (H_2CO_3) either serves as an extra source of H^+ or is reduced directly according to (14) and (15):

$$2H^+ + 2e^- \longrightarrow H_2(g) \tag{14}$$

$$2H_2CO_3 + 2e^- \longrightarrow H_2(g) + 2HCO_3^- \tag{15}$$

The dissolved iron concentration will increase until Fe^{2+} is the same as the precipitation rate of $FeCO_3$ [30]. When Fe^{2+} is released in the corrosion process, the double amount of bicarbonate forms according to (16):

$$Fe + 2H_2CO_3 \longrightarrow Fe^{2+} + H_2 + 2HCO_3^- \tag{16}$$

The pH increases until bicarbonate and carbonate becomes so high that solid $FeCO_3$ precipitates [30] as shown in equation (17):

$$Fe^{2+} + 2HCO_3^- \longrightarrow FeCO_3(s) + H_2CO_3 \tag{17}$$

When all the ferrous ions produced by corrosion precipitates as iron carbonate ($FeCO_3$), the pH remains constant and the overall reaction becomes as the state in (13).

In order to control the rate of corrosion on the pipeline, there should be passivity. Passivity is the condition existing on a metal surface because of the presence of protective film. When protective film is formed on the metal surface, it forms a coat which hinders further corrosion action on the material [26, 31]. The structure of the passive film (magnetite) formed on low carbon steel oxidizes in high temperature and has two distinct layers on the steel. The inner layer is compact and adheres well to the steel and has uniform thickness. The outer layer is a porous mass of individual crystal that would flake off the steel in some place and very nonuniform in thickness (Figure 5). This protective film is removed from the surface of the pipeline through erosion, dissolution, and turbulence resulting in more corrosion. The possible mechanisms resulting in the removal of the protective film are a follows:(i)Dissolution or removal of protective layer by hydrodynamic shear stress occurs when the shear stress is greater than the bonding force between the film and the substrate. This is a function of a mechanical process of erosion caused by the multiphase flow regime in pipeline [19, 32].(ii)In distributed flow condition, local near wall density of turbulence helps to remove the protective film. This disruption to the mass transfer boundary layer results in an enhanced corrosion rate [32, 33].(iii)Dissolution of film which is controlled by mass transfer. Thus the breakaway velocity may reflect conditions where the dissolution rate of the film is greater than the growth rate of the film [34].

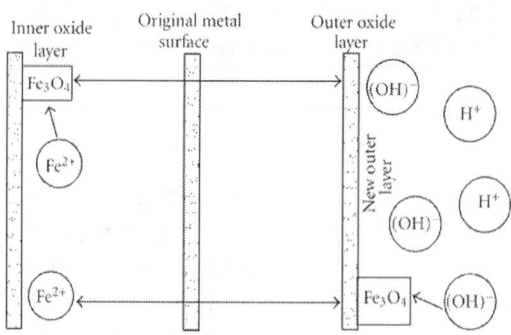

Figure 5: Schematic of magnetite double layer showing oxide formation locations.

The breakdown of protective film leads to the formation of localized corrosion that results in some of the major sources of corrosion failures like pitting, crevice, intergranular, and stress corrosion [12, 26, 35, 36]. The predominant breakdown processes are electrochemical and mechanical. Mechanical breakdown occurs when the protective film is ruptured as a result of stress or abrasive wear while, the electrochemical breakdown is a function of chemical reaction between the fluid constituent and the steel.

Types of Corrosion in Oil and Gas Pipeline

The primary chemical components that cause corrosion reaction to occur in pipeline are oxygen, acidic sulphur, and acidic chloride that dissolves in the water in the pipeline. The mechanism present in a given piping system varies according to the fluid composition, service location, geometry, temperature and so forth. In all cases of corrosion, the electrolyte must be present for the reaction to occur.

Internal Corrosion

Internal corrosion has become an increasing problem in most oil and gas pipelines as water cuts have increased and previously oil wet pipe surfaces have become water wet (providing the electrolyte for the corrosion cell) and as bacterial activities increases in the production system. Internal corrosion is the largest cause of pipeline failure in oil and gas industries [24] through different forms of corrosion like microbiologically influenced corrosion (MIC), erosion (flow enhanced) corrosion, under deposit (concentration cell) corrosion and so forth.

Erosion-Corrosion

The erosion-corrosion mechanism increases corrosion reaction rate by continuously removing the passive layer of corrosion products from the wall of the pipe. The passive layer is a thin film of corrosion product that actually serves to stabilize the corrosion reaction and slow it down. As a result of the turbulence and high shear stress in the line, this passive layer can be removed causing the corrosion rate to increase [37]. The erosion-corrosion is always experienced where there is high turbulence flow regime with significantly higher rate of corrosion than just corrosion or erosion in pipeline [38]. In a multiphase flow regime with a fully developed turbulent flow, bubbles development and collapse have been attributed to changes in mass transfer coefficient and an eventual increase in CO_2 corrosion in pipeline [34].

Under Deposit Corrosion

The under deposit mechanism can increase the corrosion reaction rate by causing a localized chemical concentration which results in pitting of the metal surface under solid deposits. These deposits appear to be composed of a corrosion product matrix with entrapment of formation solids, sand, and iron sulphide. The rate of corrosion under this mechanism is significantly lower than erosion-corrosion mechanism.

Microbiologically Induced Corrosion (MIC)

This type of corrosion is caused by bacterial activities. The bacteria produce waste products like CO2, H2S and organic acids that corrode the pipes by increasing the toxicity of the flowing fluid in the pipeline. Some bacteria like sulphate removing bacteria (SRB) consume hydrogen that is a product in a standard corrosion reaction process. This activity causes the existing corrosion rate to increase in an attempt to reach reaction equilibrium by replacing the hydrogen consumed by bacteria. Bacteria also accumulate on the pipe walls, creating deposits and under deposit corrosion. MIC is recognized by the appearance of black slimy waste material or nodules on the pipe surface as well as pitting of the pipe wall underneath these deposits.

Pitting Corrosion

Pitting is classified as a localized attack that results in rapid penetration and removal of metal at small discrete area. The initiation of a pit occurs when electrochemical or chemical breakdown exposes a small local site on a metal surface to damaging species such as chloride ion. The site where pitting occurs is where there is an environmental variation in comparison to the entire metal surface. The combination of chlorine with H_2S results in localized pitting on steel [35]. This area of pitting which is usually the anode normally get highly degraded due to enormous electron transfer between the entire large area of the metal surface which is the cathode and small anode (the pitting site).

Crevice Corrosion

Crevice corrosion results when a portion of a metal surface is shielded in such a way that the shielded portion has limited access to the surrounding environment. Such surrounding environment contain, damaging corrosion species usually chloride ion. A typical example of crevice corrosion is the crevice found at the area between two metal surfaces in close contact with a gasket or another metal surface. The environment that eventually forms in the crevice is similar to that formed under the precipitated corrosion that covers a pit. An electrochemical corrosion cell is formed from the couple between the unshielded surface and the crevice interior exposed to an environment with a lower oxygen concentration compared with the surrounding medium. The concentration of being the anode of a corrosion cell and existing in an acidic, high-chloride environment where repassivation is difficult makes the crevice interior subject of corrosion attack.

Stress Corrosion Cracking (SCC)

Stress corrosion cracking (SCC) is a form of localized corrosion which produces cracks in metals by simultaneous action of a corrodent and tensile stress. It propagates over a range of velocities from 10^{-3}

⁻10 mm/h depending upon the combination of alloy and environment involved. The geometry is such that if they grow to appropriate lengths, they may reach a critical size that results in a transition from the relatively slow crack growth rate associated with stress corrosion to fast crack propagation rates associated with purely mechanical failure. This transition happens when the stress intensity, which is a function of the geometry of the component including the crack size, reaches the fracture value for the material concerned. SCC in pipeline is a type of environmentally associated cracking (EAC). This is because the crack is caused by various factors combined with the environment surrounding the pipe. The most obvious identifying characteristic of SCC in pipeline is high pH of the surrounding environment, appearance of patches, or colonies of parallel cracks on the external of the pipe [39].

Top of the Line Corrosion (TLC)

This type of corrosion occurs due to the inability of corrosion inhibitors getting to the top of the pipeline (12 o'clock) thereby exposing it to corrodents. The inhibition effect is found to be predominant at the bottom of the line (6 o'clock), 9 o'clock and 3 o'clock where the flow of the oil or gas is taking place. This exposes the top of the line to concerted attack by the agents of corrosion with a resultant failure at some point. The primary factor that affects TLC is temperature which acts on the iron carbonate film formed. The combined effect of temperature fluctuation and condensation rate exposes the iron carbonate film to deterioration and consequently more corrosion. A study of the influence of gas flow rate on TLC shows that higher flow rate (which results in higher condensation rate) brings about more corrosion [40], while at a certain critical condensation rate, temperature and pH, TLC does not occur in gas pipelines [41]. The presence of acetic acid (HAc) has been found to enhance CO_2 TLC on carbon steel pipe, though at certain concentration level, HAc does not affect CO2 TLC in carbon steel [42].

External Corrosion

External corrosion is caused by water penetrating the insulation system and is trapped between the insulation and the external pipe wall. The corrosion cell is fuelled by the continual supply of water and oxygen from the external sources. The main area where external corrosion is found is at the field applied weld insulation packs, but it can also be at any location where the galvanized insulation jacket has been punctured or torn. Weld pack insulations that are not well sealed allow water ingress making the weld packs to be wet. A fairly high temperature is needed to drive the corrosion mechanism, and the longer the mechanism has been active, the worst the damage will be. Therefore, the hottest and coldest lines in the field should have the highest likelihood for having an external corrosion problem.

CORROSION MANAGEMENT TECH-NIQUES

Corrosion management is that part of the overall management system, which is concerned with the development, implementation, review, and maintenance of corrosion policy [7]. The corrosion policy, however, is a framework on which decision concerning corrosion issue in an industrial setting is based. This framework provides basic measures for risk determination via development of absolute risk control measures through planning, implementation, and control strategies. Corrosion management contributes to numerous benefits like statutory or corporate compliance with safety, health and environment policies, reduction in leaks, increased plant availability, reduced unplanned maintenance, and reduction in deferment costs [43]. To manage corrosion involves the utilization of a framework that will model the organization's policy through organizing, planning and implementing, measuring and reviewing, and auditing performance at all levels of execution as shown in Figure 6.

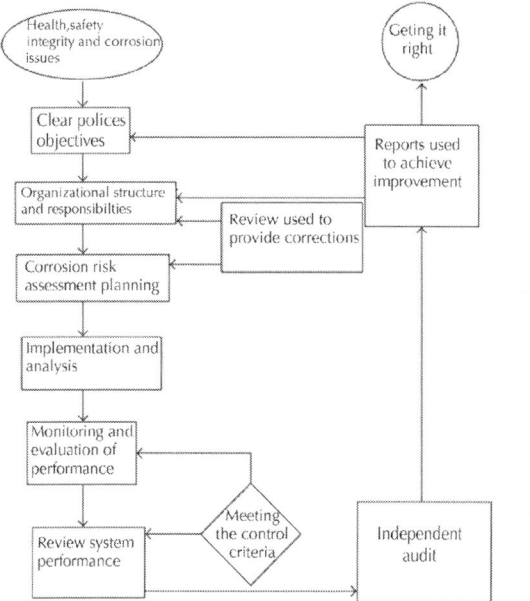

Figure 6: Corrossion management framework.

Corrosion Risk Assessment (CRA)

In planning for corrosion management, there is need for a formal identification of facilities that have the risk of degradation due to corrosion. The purpose of corrosion risk assessment is to rank facilities in order of their proneness to corrosion, identify options to remove, mitigate, or manage the risks. In order to manage corrosion risks, monitoring and inspection program will be incorporated in the overall activity schedule of an organization. The probability of failure is estimated based on the type of corrosion damage expected to occur on the component while the consequences of failure are measured against the impact of such a failure evaluated against a number of criteria. The criteria could include potential hazards to environment, risks associated with safety and integrity, or risk due to corrosion or inadequate corrosion mitigation procedure.

Typical of the risk-based procedure is the Failure Mode, Effect and Critical Analysis (FMECA) that ranks perceived risks in order of seriousness as shown in (18):

Criticality (Risk)

$$= \text{Effect (Consequences)} * \text{Mode (Probable Frequency)}, \quad (18)$$

where: failure Criticality is potential failures as examined in order to predict the severity of each failure effect in terms of safety, decreased performance, total loss of function and environmental hazards. Failure effect is potential failures assessed to determine the probable effect on process performance and the effects of the components on each other. Failure mode is the anticipated operational conditions used to identify most probable failure mode, the damage mechanism and likely locations.

Corrosion risk is the product of the probabilities of a corrosion-related failure and the consequences of such a failure [44]. The risk analysis of a pipeline is a measure of the probability of failure. The acceptable annual failure probability is dependent on the safety class [45] as shown in Table 1.

Table 1: Safety class and target annual failure probability.

Safety class	Annual failure probability
High	$<10^{-3}$
Medium	$<10^{-4}$
Low	$<10^{-5}$

Corrosion risk assessment can be carried out on a group of components which are constructed from the same material and subject to the same operating condition or an individual component. In oil and gas pipelines, the risk is analysed as either external or internal corrosion threat or environmental and operational threat. The remaining life of the pipeline is estimated against some established operational standards, while the rate of corrosion is correlated with the operating parameters of the oil and gas like CO_2, H_2S, temperature, pressure, flow rate, water cut and so forth. For effective corrosion assessment, the information concerning the operating condition of a facility will be maintained throughout the life cycle. The information is useful in formulating a corrosion risk assessment model that will be validated and modified with new assumptions overtime. For a non stable process condition, detailed re-assessment will be required at least annually but a stable process with good historical data trend will need revalidation less frequently [7].

Risk Based Inspection (RBI)

In managing oil and gas pipelines against corrosion, RBI technique is used to develop an optimum plan for the execution of the inspection activities. RBI uses findings from corrosion risk assessment (CRA) or other risk analysis to plan physical inspection procedures. A risk-based approach to inspection planning will ensure that risk is reduced to as low as reasonably practicable. It will also optimize inspection schedule, focus effort on the most critical area, and identify the most appropriate methods of inspection [46]. Planning a risk-based analysis involves listing activities, task and other elements of a project, identifying the technical risks, develop a risk ranking factor scale for each activity, document results and identify potential risks reduction actions for evaluation by personnel [47].

Corrosion Monitoring

Corrosion inspection and monitoring are key activities in ensuring, pipelines integrity are maintained and corrosion mitigated [48]. The choice of corrosion control measure is a function of fluid composition, pressure, temperature, aqueous fluid corrosivity, facility, and technical culture inherent in an establishment. In monitoring and inspection of pipelines, data are collected to enhance corrosion control by way of predicting the remaining life and the suggestion of possible mitigation measures that will help to enhance serviceability will largely depend on the experience of the personnel. A thorough practice for corrosion management involves the monitoring of corrosion risks through proactive and reactive monitoring techniques. In management of pipeline corrosion in oil and gas industries, proactive technique which involves determination of the corrosion standpoint prior to failure is utilized. This involves in-line and on-line monitoring system. In this system, data which could enhance the knowledge of the rate of corrosion degradation are collected and steps are taken to prevent failure. In-line system cover the installation of devices directly into the pipeline like corrosion coupons, biostuds and so forth. These need to be extracted for analysis periodically. On-line monitoring techniques include deployment of corrosion monitoring devices either directly into the process or fixed permanently to the facility. These include electrical resistance (ER) probes, linear polarization resistance (LPR) probes, fixed ultrasonic (UT) probes, acoustic emission and so forth.

Whereas some corrosion monitoring techniques can be used for continuous monitoring, others are used for periodic monitoring. Corrosion monitoring techniques can either be direct or indirect parameter measure. This is summarized in Table 2.

Table 2: Summary of corrosion monitoring techniques

Direct method	Indirect method
Non-destructive inspection (NDI)	Biological counts
Material test coupons	Hydrogen probes
Electrical resistance (ER) probes	pH probes
Linear polarization resistance (LPR)	Specific ions
Elector-chemical impedance spectroscopy (EIS)	Temperature
Electro-chemical noise (EN)	conductivity
Galvanic current (GC)	Electrical potential monitor

Corrosion Mitigation Strategies

After corrosion risk assessment and data collection and analysis are completed, there is need for corrective action on the facility; this depends on the level of the deterioration experienced by facility. The approaches available for mitigating corrosion in pipeline includes, coating surfaces to act as a barrier or perhaps provide sacrificial protection, the addition of chemical specie to the environment to limit corrosion, alteration of alloy chemistry to make it more resistance to corrosion and utilization of alternative material [24].

Effective corrosion mitigation involves a good approach to assessment linked to inspection monitoring during initial design and re-evaluation of pipeline with respect to the selection of inhibitors. The summary of inhibitor selection for carbon steel pipeline at different risk categories is shown in Table 4.

Corrosion can be prevented or controlled by understanding the principle underlying corrosion process. This understanding has been the basis for the development of a number of corrosion prevention measures. The basic corrosion control measures are based on electrochemical driving force as shown in Pourbaix diagram in Figure 7. Table 3 shows the different pipeline corrosion mitigation strategies.

Table 3: Shows the different pipeline corrosion mitigation strategies

Mitigation strategy	Option	Remarks
Appropriate materials	Use of corrosion resistant alloys, non-metallic materials like Reinforced composite, thermoplastic-lined and polyethylene pipelines. Consider use of internally coated carbon steel pipeline systems (i.e., nylon or epoxy coated) with an engineered joining system.	(i) Non-metallic materials may be used as a liner or a free standing pipeline depending on the service conditions. (ii) Selection of appropriate material at construction and major refurbishment stage is necessary.
Chemical treatment	Corrosion inhibitors, biocides, oxygen scavengers, gas blanketing, vacuum deaeration	(i) The presence of small amounts of oxygen (parts per billion) or bacteria will accelerate corrosion. (ii) Provides a barrier between corrosive elements and the pipe surface
Coating and lining	Organic Coatings, metallic coatings, lining, cladding	Useful for internal and external corrosion prevention
Cathodic protection	Sacrificial anodes, impressed current systems, hybrid system	Need ability to monitor performance on-line.
Process control	Identify key parameters: pH, temperature, pressure, Flow rate, water chemistry, pH, chlorides, dissolved metals, bacteria, suspended solids, chlorine, oxygen, and chemical residuals	(i) Changes in operating conditions will influence the corrosion potential. Production information can be used to assess corrosion susceptibility based on fluid velocity and corrosivity (ii) Trends in dissolved metal concentration (i.e., Fe, Mn) can indicate changes in corrosion activity
Design detailing	Ensure ease of access and replacement: (i) Install valves that allow for effective isolation of pipeline segments from the rest of the system (ii) Install binds for effective isolation of in-active pipeline segments	Allows the effective suspension and discontinuation of pipeline segments: (i) Removes potential "deadlegs" from the gathering system (ii) Develop shut-in guidelines for the timing of required steps to isolate and lay up pipelines in each system

Table 4: Corrosion inhibitor risk categories

Risk category	Max inhibitor availability	Max expected uninhibited corrosion rate (mm/yr)	comments	Proposed category name
1	0%	0.4	Benign Fluid, corrosion inhibitor use not anticipated. Predicted metal loss accommodated by corrosion allowance	Benign
2	50%	0.7	Corrosion inhibitor probably required but with expected corrosion rates there will time be time to review the need for inhibition based on inspection data.	Low
3	90%	3	Corrosion inhibition required for majority of field life but inhibitor facilities need not be available from day one.	Medium
4	95%	6	High reliance on inhibition for operational life time. Inhibitor facilities most be available from day one to ensure success	High
5	>95%	>6	Carbon steel and inhibition is unlikely to provide integrity for full field life. Select corrosion resistant material or plan for repair and replacement	Unacceptable

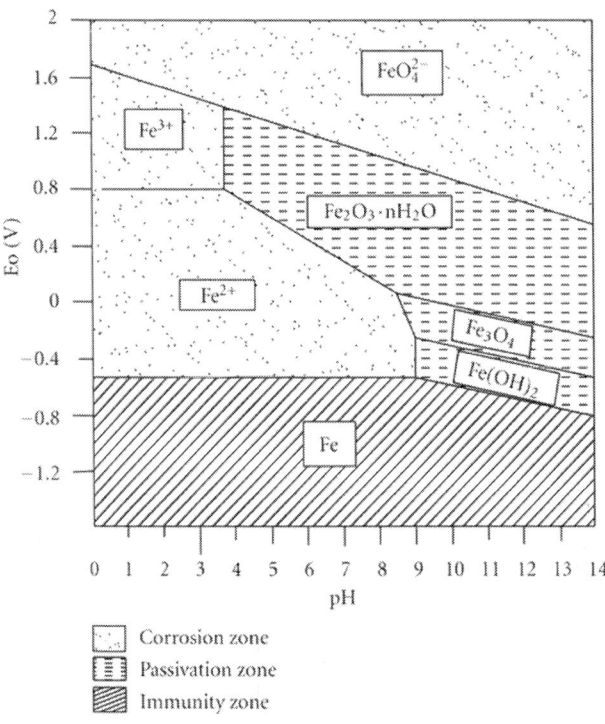

Figure 7: Pourbaix diagram of iron corrosion, passivity and immunity.

CONCLUSIONS

The prevalence of corrosion in oil and gas industry has resulted in enormous investment in technology to help combat impacts like loss of containment, leakages, death of personnel an environmental pollution. To this end, new oil and gas fields are developed using experiences generated from previous fields with similar characteristics. Efforts of design personnel at ensuring that the carbon steel materials are operating in the environment of immunity or passivity as shown in Pourbaix diagram (Figure 7) are yielding results via introduction of high corrosion resistant carbon alloys for pipelines. In other instances, specialized corrosion resistant materials have been used for lining the pipelines while reinforced composite and PVC materials have been utilized as alternative material for pipeline construction.

Advancements in inspection and monitoring techniques are also aiding corrosion experts in decision concerning the "when" and "how" pipelines are managed in a bid to optimize performance and cost. The proliferation of different empirical, statistical, and mechanistic prediction models for corrosion prediction is aiding personnel in managing the integrity of the pipelines through different mitigation strategies.

Finally, if pipeline corrosion which is a major contributor to nonproductive time (NPT) in oil and gas production will be reduced to the barest minimum, a corrosion management policy with a well-defined structure that includes responsibilities, reporting routes, practices, procedures, and resources has to be strictly followed in the oil and gas industries. The effectiveness of the policy will therefore depend on the willing of the leadership and commitment of other personnel at all ranks.

REFERENCES

1. R. Nicholson, J. Feblowitz, C. Madden, and R. Bigliani, "The Role of Predictive Analytics in Asset Optimization for the Oil and Gas Industry-White Paper," 2010, http://www.tessella .com/wp-content/uploads/2008/02/IDCWP31SA4Web.pdf.

2. J. Neelamkavil, "A review of Existing Tools and Their Applicability to Maintenance Management. Report # RR-285," http:// pdf.aminer.org/000/274/575/a decision support system for transmission facility maintenance.pdf.

3. Oil and Gas Enhanced Production Services Industry to 2016— Enhanced Oil Recovery (EOR) Driving E&P Activity in Depleting Hydrocarbon Reservoirs, http://www.reportlinker .com/ p0845623/Oil-and-Gas-Enhanced-Production-ServicesIndustry-to-Enhanced-Oil-Recovery-EOR-Driving-E-P-Activity-in-Depleting-Hydrocarbon-Reservoirs.html.

4. Control of Major Accident Hazards, "Ageing Plant Operational Delivery Guide," http://www.hse.gov.uk/comah/guidance/ageing-plant-core.pdf.

5. P. Horrocks, D. Mansfield, K. Parker, J. Thomson, T. Atkinson, and J. Worsley, "Managing Ageing Plant," http://www.hse .gov.uk/ research/rrpdf/rr823-summary-guide.pdf.

6. "Cost of Corrosion to Exceed $1 Trillion in the United States in 2012—G2MT Labs -The Future of Materials Condition Assessment," http://www.g2mtlabs.com/2011/06/nace-cost-of -corrosion-study-update/.

7. Review of Corrosion Management for Offshore Oil and Gas Processing, HSE OffshoreTechnology Report 2001/044, 2001.

8. Corrosion in the Oil Industry(Oilfield review) Schlumberger, http://www.slb.com/resources/publications/industry   articles/ oilfield review/1994/or19940401 corrosion.aspx.

9. B. Khajota, D. Sormaz, and S. Nesic, "Case-based reasoning model of CO2 corrosion based on field data," CORROSION, 2007, paper no. 07553.

10. K. U. Raju, "Successful scale mitigation strategies in Saudi Arabian oil fields," in International Symposium on Oilfield Chemistry, The Woodlands, Tex, USA, April 2009, paper no. 121679.

11. A. Darwin, K. Annadorai, and K. Heidersbach, "Prevention of corrosion in carbon steel pipeline containing hydrostatic water-an overview," in CORROSION, March 2010, paper no. 10401.

12. J. Wen, T. Gu, and S. Nesic, "Investigation of the effects of fluid flow on SRB biofilm," in CORROSION, 2007, Paper no. 07516.

13. B. Hedges, H. J. Chen, T. H. Bieri, and K. Sprague, "A review of monitoring and inspection technique for CO2 and H2S corrosion in oil and gas production facilities: location, location, location," in CORROSION, 2006, paper no. 06120.

14. M. Singer, B. Brown, A. Camacho, and S. Nesi˘ c, "Combined effect of carbon dioxide, hydrogen sulfide, and acetic acid on bottom-of-the-line corrosion," Corrosion, vol. 67, no. 1, 2011.

15. K. L. J. Lee and S. Nesic, "EIS investigation of CO2/H2S corrosion," in CORROSION, April 2004, paper no. 04728.

16. K. D. Ralston and N. Birbilis, "Effect of grain size on corrosion: a review," Corrosion, vol. 66, no. 7, pp. 0750051–07500513, 2010.

17. Y. Song, A. Palencsar, G. Svenningsen, J. Kvarekv ´ al, and T. ° Hemmingsen, "Effect of O2 and temperature on sour corrosion," Corrosion, vol. 68, no. 7, pp. 662–671, 2012.

18. A. Kale, B. H. Thacker, N. Sridhar, and C. J. Waldhart, "A probabilistic model for internal corrosion of gas pipelines," in

Proceedings of the 5th Biennial International Pipeline Conference (IPC '04), pp. 2437–2445, Calgary, Canada, October 2004. 10 ISRN Corrosion

19. S. Nesic, J. Cai, and K.-L. J. Lee, "A multiphase flow and internal corrosion prediction model for mild steel pipeline," in CORROSION, 2005, Paper no. 05556.

20. W. Sun and S. Nesic, "A mechanistic model of H2S corrosion of mild steel," in CORROSION, 2007, paper no. 07655.

21. X. Hu, V. D. Souza, A. Neville, and J. Well, "Prediction of erosion-corrosion in oil and gas- a systematic approach," in CORROSION, 2008, paper no. 08540.

22. X. Tang, C. Li, F. Ayello, J. Cai, and S. Nesic, "Effects of oil type on phase wetting transition and corrosion in oil-water flow," in CORROSION, NACE International, 2007, Paper no. 017170.

23. Y. Xian and S. Nesic, "A stochastic prediction model of localized CO2 corrosion," in CORROSION, 2005, paper no. 05057.

24. CAPP, "Best Management Practices: Mitigation of Internal Corrosion in Oil Effluent Pipeline Systems," 2009, http:// www. capp.ca/getdoc.aspx?DocId=155641&DT=PDF.

25. B. A. Shaw and R. G. Kelly, "What is corrosion?" Electrochemical Society Interface, vol. 15, no. 1, pp. 24–26, 2006.

26. J. Kruger, "Electrochemistry of Corrosion," 2001, http:// electrochem.cwru.edu/encycl/art-c02-corrosion.htm.

27. V. Fajardo, C. Canto, B. Brown, and S. Nesic, "Effect of organic acids in CO2 corrosion," in Proceedings of the NACE International Conference and Exposition CORROSION, 2007, paper no. 07319.

28. P. S. Joshi, G. Venkateswaran, K. S. Venkateswarlu, and K. A. Rao, "Stimulated decomposition of Fe(OH)2 in the presence of AVT chemicals and metallic surfaces—relevance to lowtemperature feedwater line corrosion," CORROSION, vol. 49, no. 4, pp. 300–309, 1993.

29. R. Nyborg, "Controlling internal corrosion in oil and gas pipeline," Business Briefing-Exploration & Production: The Oil & Gas Review, no. 2, pp. 70–74, 2005.

30. A. Dugstad, E. Gulbrandsen, M. Seiersten, J. Kvarekval, and R. Nyborg, "Corrosion testing in multiphase flow, challenges and limitations," in CORROSION, 2006, paper no 06598.

31. R. N. Kig, "A review of fatigue crack growth in air and seawater," Offshore Technology Report OTH96 511, HSE, 1996.

32. A. Keating and S. Nesic, "Prediction of two-phase erosioncorrosion in bends," in Proceedings of the 2nd International Conference on CFD in Minerals and Processes Industries CSIRO, Melbourne, Australia, December 1999.

33. S. Nesic and J. Postlethwaite, "Relationship between the structure of disturbed flow and erosion-corrosion," Corrosion, vol. 46, no. 11, pp. 874–880, 1990.

34. H. Wang, W. Paul Jepson, J.-Y. Cai, and M. Gopal, "Effect of bubbles on mass transfer in multiphase flow," in CORROSION, 2000, paper no. 00050.

35. H. Fang, B. Brown, and S. Nesiæ, "E ˘ ffects of sodium chloride concentration on mild steel corrosion in slightly sour environments," in CORROSION, vol. 67, no. 1, January 2011.

36. E. Mysara Mohyaldinn, N. Elkhatib, and C. Mokhtar Ismail, "A computational tool for erosion/corrosion prediction in Oil/ Gas production facilities," in Proceedings of 3rd International Conference on Solid State Science & Technology (ICSSST '10), Kuching, Malaysia, December 2010.

37. Sh. Hassani, K. P. Roberts, S. A. Shirazi, J. R. Shadley, E. F. Rybicki, and C. Joia, "Flow loop study of NaCl concentration effect on erosion, corrosion, and erosion-corrosion of carbon steel in CO2-saturated systems," in CORROSION, vol. 68, no. 2, February 2012.

38. A. A. Sami and A. A. Mohammed, "Study synergy effect on erosion-corrosion in oil and gas pipelines," Engineering and Technology, vol. 26, no. 9, 2008.

39. M. Baker Jr., "Stress Corrosion Cracking Study," 2004, http:// www.polyguardproducts.com/products/pipeline/TechReference/ SCC Report-Final Report with Database.pdf.

40. S. Olsen and A. Dugstad, "Corrosion under dewing conditions," in CORROSION, 1991, paper no. 472.

41. F. Vista and K. Alam, "Semi-empirical model for prediction of top-of-the-line corrosion risk," in CORROSION, 2002, paper no. 02245.

42. C. Mendex, M. Singer, A. Camacho, S. Hernndez, and S. Nesic, "Effect of acetic acid pH and MEG on CO2 top of the line corrosion," in CORROSION, 2005, paper no. 05278.

43. D. Storey, "A Service Company's Experience with Pipeline Integrity Management," 2004, http://www.roseninspection .net/ MA/papers/2004-11 PipelineIntegrityManagement.pdf.

44. P. O. Gartland and J. Roy, "Application of internal corrosion modelling in risk assessment of pipeline," in CORROSION, 2003, paper no. 03179.

45. Det Norske Veritas (DNV RPG 101), Recommended Practice DNV-RP-101: Corroded Pipelines, 2010.

46. Det Norske Veritas (DNV RPG 101), "Risk Based Inspection of Topsides Static Mechanical Equipment, 2001".

47. P. K. John and L. D. John, "Risk factor analysis—a new qualitative risk management tool," in Proceedings of the Project Management Institute Annual Seminar & Symposium, September 2000.

48. E. J. Carl, A. B. John, and G. T. Neil, "Improving plant reliability through corrosion monitoring," in Proceedings of the Process Plant Reliability, Houston, Tex, USA, November 1995.

Chapter 5

Temporal Variations of the Chemical Composition of Three Seaweeds in Two Tropical Coastal Environments

Andyara Nascimento[1],
Carina Coelho-Gomes[2], Elisabete Barbarino[1, 3],
and Sergio Oliveira Lourenço[1]

[1]Department of Marine Biology, Fluminense Federal University, Niterói, Brazil

[2]Maria Thereza College, Niterói, Brazil

[3]Post-Graduate Programme in Chemistry, Fluminense Federal University, Niterói, Brazil

ABSTRACT

The seaweeds Chaetomorpha antennina, Gymnogongrus griffithsiae and Ulva fasciata were studied regarding tissue concentrations of total nitrogen, total phosphorus, total protein, hydrosoluble protein, total carbohydrate, chlorophyll a and total carotenoid throughout a 39-month survey in two coastal environments of Rio de Janeiro State, Brazil. One of the sites (Itapuca Stone) has high concentrations of dissolved nutrients and an intense long-term process of cultural eutrophication; the second site (Bananal Inlet) is thought to have lower concentrations of dissolved nutrients and no relevant anthropic impact. Seaweeds experienced changes in the concentrations of the substances in the thalli; however they did not show any cyclic seasonal pattern, except for pigments, with lower values in summer in both sites. The differences found for each species in each sampling at the sites were small (e.g. U. fasciata, more total nitrogen at Itapuca Stone) or absent (e.g. C. antennina, no significant differences for hydrosoluble protein in the sites). Differences in the concentrations of dissolved nutrients in the sites did not generate contrasting chemical profiles in the seaweeds. There is no evidence of nitrogenor phosphorus-limitation in any season. It is presumable that the concentrations of dissolved nutrients at the nutrient-poorer site are sufficient to generate high concentrations of the substances in the thalli of the species tested, similar to the concentrations measured in the eutrophic site. Experimental data are needed to elucidate the factors that promote the success of the species tested under contrasting nutrient availability and environmental disturbance.

INTRODUCTION

Growth of macrophytes and phytoplankton in tropical coastal waters is generally limited by nutrient availability [1] . Human use of coastal areas has greatly increased the inputs of nitrogen and phosphorus into many aquatic systems, with resultant impacts at the population and ecosystem level [2] . Increased abundance of nuisance macroalgae is among the direct consequences of nutrient loading [3] .

Studies on the abundance of opportunistic seaweeds and measurements of dissolved nutrients are traditional approaches used to add information to evaluate the trophic state of a given ecosystem. However, other parameters may also be used to assess some ecological characteristics of coastal environments. For instance, monitoring the concentration of total N and P in macroalgal tissues may be a more useful indicator of enrichment or eutrophication potential [4] , since total nutrient concentration in the algal tissue integrates the nutrient regime over time [5] [6] .

In addition to measurements of tissue N and P, other chemical parameters can be useful in this field. Analyses of protein, carbohydrate and photosynthetic pigments can aggregate more information for the understanding of the behavior of algal species as responses to environmental conditions. Protein in the thalli is mainly influenced by nitrogen availability [7] . Both experimental and field studies have demonstrated that seaweeds tend to accumulate higher concentrations of protein and chlorophyll when dissolved nitrogen is available in high concentrations [8] . The values in the thalli tend to be relatively higher in specimens living in eutrophic environments or those that have been previously submitted to high concentrations of dissolved nutrients in a period before sampling [9] .

High concentrations of carbohydrate in contrast with low concentrations of protein are frequently related to nitrogen deficiency in algae [10] . Under long-term short supply of nitrogen, it is observed that an increase in total carbohydrate and a progressive decrease of the concentration of nitrogenous substances (protein, pigments, intracellular inorganic nitrogen, nucleic acids, etc.) over time occur, and this is a universal behavior of seaweeds [7] [11] and microalgae [12] . Nitrogen-bearing substances may be partially consumed as alternative sources of nitrogen by algal species under nitrogen starvation [7] [12] .

Data on pigment composition are also important to assess responses to environmental factors, such as temperature, salinity, dissolved nutrients and irradiation. The pigment content may increase in response to the environmental factors such as high nutrient availability [13] or decrease as a consequence of excess of solar radiation and exposure to UV radiation [14] . Damage caused by UV radiation may be especially relevant in tropical environments, where seaweeds are particularly exposed to high irradiation [14] [15] .

Studies on tissue chemical composition of macroalgae are predominantly carried out in temperate environments [6] [13] [16] -[21] . By comparison, information on tissue chemical composition of algae from tropical and subtropical environments is relatively scarce [22] -[27] , and more data are needed from the tropics.

In this study we report on the temporal variations of tissue N, P, N:P atomic ratio, protein, carbohydrate, and photosynthetic pigments (chlorophyll and carotenoids) of the green algae Chaetomorpha antennina and Ulva fasciata and the red alga Gymnogongrus griffithsiae. The three macroalgal species are common in two tropical sites of Rio de Janeiro State, Brazil, with different trophic states: Bananal Inlet (oligotrophic-mesotrophic) and Itapuca Stone (eutrophic-hypereutrophic). Comparisons were made between algal substances and the concentrations of dissolved nutrients in the systems in this 3-year assessment to evaluate the effects of excess of nutrients on the chemical composition of the species studied.

MATERIALS AND METHODS

Sampling Sites

Both sampling sites are located in Niterói municipality, State of Rio de Janeiro, Brazil (Figure 1). Bananal Inlet (23°58›S, 43°01›W) corresponds to the marine part of an environmental protected area (Serra da Tiririca State Park), with restricted access for recreational uses.

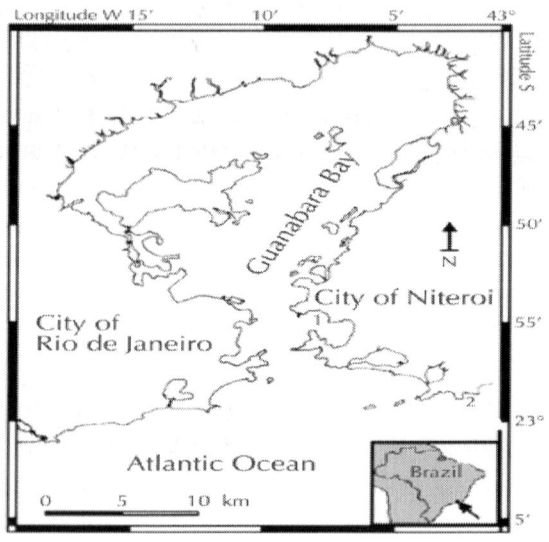

Figure 1: Sampling sites in Rio de Janeiro State, Brazil. 1 = Itapuca Stone. 2 = Bananal Inlet.

The area is not inhabited, but human occupation can be seen close to the limits of the park (5 - 6 km away from the sampling site). The terrestrial part of the park is a mountain area covered by a tropical rain forest (Atlantic Forest). This site is considered protected from relevant human impacts. Macroalgal floristic studies are still scarce in this site, but preliminary results indicate the existence of 92 species at the intertidal zone (Moreira, unpublished data).

The second site is Itapuca Stone (23°04›S, 43°08›W), located in Guanabara Bay. The site is in the urban area of Niterói City, and it is located near the entrance of the Bay (Figure 1), which promotes a local dilution in the typical high levels of pollution of this coastal system and a faster water turnover [28] . Inner areas of Guanabara Bay show a low water exchange rate [29] due to geomorphological features and human occupation of coastal areas. The Bay comprises an area of 381 km^2 and an estimated 2-billion m^3 of water. Its hydrographic basin (4000 km^2) includes 35 rivers that contribute substantially to the freshwater input. Guanabara Bay is considered a eutrophic or hypereutrophic environment (depending on the specific part of the Bay), highly disturbed by anthropic impacts [30] .

Considering the environmental characteristics described here, we hypothesized that the seaweeds of Itapuca Stone (Guanabara Bay) would present permanently high concentrations of tissue N and P; in addition, the seaweeds in Itapuca Stone would not show significant variations in their tissue substances throughout the year and no inter-annual changes in the chemical substances analyzed. On the other hand, temporal changes in algal tissue substances would be expected for the seaweeds of Bananal Inlet.

Seaweeds

Three macroalgal species were analyzed, and their identification was carried out following a widely accepted checklist [31] . Chlorophyta: Chaetomorpha antennina (Bory) Kützing and Ulva fasciata L. Rhodophyta: Gymnogongrus griffthsiae (Turner) Mart.

Sampling

Sampling began in December 2000 (end of the austral spring) and continued through February 2004 (austral summer). Samples were collected every 30 - 75 days, depending on the tidal regime and season. Samples were collected in the intertidal area only.

Whole thalli of adult plants were collected in early morning and washed in the field with seawater to remove epiphytes, sediment and detritus. At least 15 whole plants of each species were collected, independent of the size of each seaweed. All species were typically found at the same specific points in the site throughout the study (e.g. C. antennina was sampled always at the same rocks). The plants were placed in plastic bags, and kept on ice until return to the laboratory (less than one hour). In the laboratory, samples were gently brushed under running seawater, rinsed with distilled water, and dried at 60°C for at least three days and until constant weight, to determine the percentage of moisture in the tissues. The dried material was ground into a powder and kept in desiccators containing silica-gel at room temperature until N and P tissue analyses.

Samples for pigment were analyzed immediately after the preparation of the algal material, in the same day of the field trip, using wet thalli. Samples for protein and carbohydrate were cleaned, weighted (wet weight) and stored at 4°C until analyses, up to five days later. At the time of each collection of macroalgae, four 250 ml-water samples (n = 4) for dissolved nutrient analysis were taken from 15 - 20 cm below the water surface, as well as measurements of local temperature at the same depth. The samples of water were filtered through cellulose membrane filters (Millipore®HAWP 0.45 μm pore) and kept at −20°C until spectrophotometric determinations of dissolved nutrients. Each sample was measured at least three times to obtain accurate results, and the results showed in this study represent mean values for four independent samples collected in the field for each sampling.

Tissue Analyses

Total N and P were determined in algal tissue after peroxymonosulphuric acid digestion, using a Hach digestor (Digesdhal$^{\circ}$, Hach Co.) [32] . Total N and P contents in the samples were determined spectrophotometrically after specific chemical reactions. For analytical details see Lourenço et al. [15] . For each species and sampling four independent (from different plants) measurements of tissue N and P were performed (n = 4).

The Lowry et al. method [33] was used to evaluate hydrosoluble protein in the samples, with bovine serum albumin as a protein standard. Spectrophotometric determinations were done at 750 nm. Results obtained for total nitrogen were used to calculate the total protein content, using the nitrogen-to-protein conversion factors proposed by Lourenço et al. [34] . Carbohydrate was extracted with 80% H_2SO_4, according to Myklestad & Haug [35] and determined spectrophotometrically at 485 nm by the phenol-sulphuric acid method [36] , using glucose as a standard. Pigment extraction was performed in methanol, at 4°C, for 20 h. Chlorophyll a and total carotenoid was determinated spectrophotometrically as described by Lorenzen [37] and Strickland & Parsons [38] , respectively.

Dissolved Nutrients

For the quantification of nutrient ions in seawater, spectrophotometric determinations of nitrate and nitrite [39] , ammonium/ammonia [40] , urea and phosphate [41] were performed, following standard procedures.

Physical and Meteorological Parameters

Salinity was measured with a hand refractometer (Shibuya Optical, model S-10) using four samples (n = 4) collected in the field in each trip. Air temperature and seawater temperatures were measured with a mercury-column thermometer (Incoterm Co., Brazil).

Meteorological data (average monthly air temperature and precipitation) were obtained from the Fluminense Federal University Meteorological Station, located in Niterói, beside Guanabara Bay.

Statistical Analysis

The results for each species separately and for total measurements of all species combined were analysed by single-factor analysis of variance

(ANOVA) with significance level α = 0.05 [42] , followed with a Tukey's multiple comparison test. Suitable transformations of data (e.g. log of the actual data) were made when necessary. Time was the only factor considered in ANOVA.

RESULTS

Table 1 shows small temperature variations throughout the study. Maximum temperatures tended to be achieved in December-February (austral summer). Maximum monthly average temperatures were obtained in December 2003 and February 2004. Similar trends were obtained for atmospheric precipitation, with higher values obtained in summer months, and a maximum record in December 2001.

Measurements of salinity were typically lower at Itapuca Stone, where they fluctuated between 29.5 and 34.9 psu throughout the study (Table 2). At Bananal Inlet, minor variations in salinity were recorded, with values fluctuating around 35 psu (except in January 2004, when 31.2 was recorded). Conversely, variations in water temperature were wider in Bananal Inlet, with a difference between maximum and minimum mean values of 9°C (Table 2), ca. three times that recorded at Itapuca Stone (3°C). Air temperatures during field trips were similar in both sampling sites, however higher variations were recorded at Bananal Inlet.

In general, higher concentrations of all dissolved nutrients were found at Itapuca Stone, although in some observations the concentrations of nutrients were similar in both sites (Table 3). Typical concentrations of ammonium/ammonia were > 5 μM at Itapuca Stone and < 2 μM at Bananal Inlet, with significant differences between the sites (p < 0.0001). In Itapuca Stone, nitrite concentrations were typically ca. three times higher than those of Bananal Inlet (p < 0.0001), and a similar trend was recorded for nitrate. At Bananal Inlet maximum values for nitrate and nitrite were found in late summer/early autumn (Table 3). Urea tended to show higher concentrations in late spring and in summer, and lower values in winter, with higher concentrations at Itapuca Stone (p = 0.0231). Total nitrogen was influenced mainly by dissolved ammonium/ammonia and nitrate, the nitrogenous ions presented in higher concentrations in both sampling sites. Higher values for total nitrogen tended to be achieved in summer at Bananal

Inlet (maximum of 12.4 µM, January 2004), and in winter at Itapuca Stone (maximum of 36.2 µM, July 2003). Typical concentrations of phosphate were ca. three times higher at Itapuca Stone than at Bananal Inlet (p < 0.0001), however the N:P atomic ratio were similar for both sites (p = 0.38), with overall fluctuations around 15:1 in seawater.

Wide variations among the three species were found for total tissue nitrogen (Figure 2). G. griffithsiae and U. fasciata tended to show higher concentrations of tissue nitrogen, while C. antennina presented lower values. In many comparisons U. fasciata showed differences for the measurements obtained in the sites, with higher values at Itapuca Stone. For the other species differences were small or not significant for monthly comparisons of the sites. C. antennina showed minor variations in tissue phosphorus throughout the study in both sites (Figure 3). A similar trend was obtained for G. griffithsiae, but in some comparisons higher values were recorded at Bananal Inlet. Among the three seaweeds, U. fasciata showed wider variations in tissue phosphorus in this study, with a trend to show higher concentrations of tissue P at Itapuca Stone.

Variations in tissue N:P ratio were wider for G. griffithsiae, varying from 10:1 (Bananal Inlet, December 2000) to 34:1 (Itapuca Stone, January 2003). For most of the comparisons, values of tissue N:P ratio were not significantly different in both sites for the three species (Figure 4).

As total protein was calculated using nitrogen-to-protein conversion factors, the same general trends described for total nitrogen were found (Figure 5). Typical values for hydrosoluble protein were higher than 15% of d.w., with U. fasciata showing percentages higher than the other species in most observations (Figure 6). Changes in hydrosoluble protein followed the same general description presented for tissue N and total protein, with U. fasciata showing more protein at Itapuca Stone, and small or null differences between the sites for the other species.

Carbohydrate was the most abundant component for all species, with typical concentrations > 40% d.w. in almost all measurements (Figure 7). G. griffithsiae showed maximum concentrations of total carbohydrate, with more than 60% of d.w. in some observations. G. griffithsiae tended to show higher concentrations of total carbohydrate in Bananal Inlet throughout the study.

Clorophyll a and total carotenoid showed wide variations in the measurements throughout the study (Figure 8 and Figure 9). C. antennina showed virtually the same concentrations of clorophyll a and total carotenoid in both sites, but G. griffithsiae tended to present higher concentrations in Bananal Inlet and U. fasciata at Itapuca Stone. For all species lower values were measured after the summer, and higher values tended to be found in autumn and winter.

Table 1: Atmospheric precipitation and air temperature collected daily at the Fluminense Federal University Meteorological Station throughout the period of this study

Month/year	Precipitation (mm)		Air temperature (˚C)		
	Average	Total	Minimum	Average	Maximum
Nov/2000	1.8	57.2	21.8	24.9	29.2
Dec/2000	3.1	94.8	22.9	26.3	31.3
Jan/2001	2.2	66.8	23.2	27.0	32.7
Feb/2001	3.4	95.6	24.2	28.3	35.0
Mar/2001	2.1	65.4	23.4	27.6	34.3
Apr/2001	0.2	4.8	22.3	26.3	32.4
May/2001	2.5	78.2	19.9	23.5	29.0
Jun/2001	2.0	56.8	19.5	23.0	28.3
Jul/2001	2.3	72.4	20.2	23.5	27.2
Aug/2001	0.2	5.4	22.5	24.8	29.0
Sep/2001	2.0	30.8	22.9	25.9	30.4
Oct/2001	1.6	49.2	23.4	26.6	31.9
Nov/2001	3.4	101.0	23.2	25.9	30.8
Dec/2001	8.1	251.8	23.7	27.7	33.8
Jan/2002	2.1	65.2	22.5	26.3	32.1
Feb/2002	4.1	114.4	21.3	24.9	30.4
Mar/2002	1.0	31.6	20.1	23.5	29.1
Apr/2002	0.1	4.4	18.9	21.9	26.3
May/2002	5.1	157.2	20.0	23.7	29.3
Jun/2002	1.9	23.2	18.9	21.9	26.2
Jul/2002	0.8	24.8	21.2	25.2	31.0
Aug/2002	0.6	18.6	22.4	25.9	31.1
Sep/2002	2.9	87.2	23.5	26.7	31.6

Oct/2002	0.9	28.8	23.9	27.1	32.5
Nov/2002	7.0	108.6	23.9	28.2	34.8
Dec/2002	5.5	171.6	23.6	27.5	33.8
Jan/2003	6.8	211.8	22.1	25.9	31.5
Feb/2003	0.1	1.8	19.2	23.0	28.6
Mar/2003	6.7	208.8	18.7	23.0	29.7
Apr/2003	1.9	56.0	17.2	21.6	27.8
May/2003	1.6	49.0	17.4	21.2	26.5
Jun/2003	1.5	45.4	18.9	22.4	27.1
Jul/2003	0.9	29.0	20.0	23.6	28.7
Aug/2003	1.0	32.0	21.8	25.3	30.3
Sep/2003	3.0	90.0	23.0	26.3	31.3
Oct/2003	4.3	132.0	22.5	25.7	30.3
Nov/2003	6.4	192.4	22.6	26.1	31.2
Dec/2003	3.8	116.8	21.8	24.9	29.2
Jan/2004	4.9	152.4	22.9	26.3	31.3
Fev/2004	4.4	127.8	23.2	27.0	32.7

Table 2: Average values of salinity and temperature measured at the sampling sites during part of the field trips. Results for salinity represent the mean values of four determinations ± the standard deviation (n = 4). Data for the first 18 months of this study are not presented

Sampling	Salinity (psu)		Seawater temperature (°C)		Air temperature (°C)	
	Itapuca Stone	Bananal Inlet	Itapuca Stone	Bananal Inlet	Itapuca Stone	Bananal Inlet
Aug/2002	33.9	34.6	24.0	22.0	23.5	21.5
Oct/2002	33.9	35.4	23.0	19.0	25.0	22.0
Nov/2002	33.7	35.2	25.0	23.5	26.7	25.0
Dec/2002	29.5	34.9	25.0	21.0	25.5	25.0
Jan/2003	30.2	34.9	26.0	23.5	28.0	25.2
Feb/2003	32.1	36.2	24.0	27.0	28.0	27.0
Mar/2003	34.9	35.1	26.0	26.0	26.5	26.0
May/2003	34.8	35.2	23.5	25.0	25.5	25.5
Jun/2003	34.7	35.0	23.0	23.0	24.0	20.5

Jul/2003	34.1	34.9	23.0	22.5	23.0	22.0
Aug/2003	32.1	34.9	23.0	22.2	22.5	21.7
Sep/2003	34.5	35.0	23.0	23.0	23.0	23.5
Nov/2003	33.7	34.1	25.5	23.0	27.0	24.5
Jan/2004	31.5	31.2	24.5	17.0	26.5	23.0
Feb/2004	29.6	35.8	25.5	18.0	26.0	23.0

DISCUSSION

Dissolved Nutrients

Results confirmed that concentrations of dissolved nutrients at Itapuca Stone were higher than at Bananal Inlet, but the differences between the sites were not intense. In some observations no statistical difference was detected between the sites, and in some monthly comparisons the absolute values measured in Bananal Inlet were only 30% - 50% lower than in Itapuca Stone.

Some hypotheses can be considered to explain the small differences in dissolved nutrients in the sites. The Inlet is the marine part of the Serra da Tiririca State Park, with most of its area comprising a rain forest on mountains. The topographical characteristics of the area possibly favor the transport of nutrients from the forest soil to the Inlet, especially after rainfalls. As a typical concentration of nitrogen in soil may be three orders of magnitude higher than that of the seawater, the run-off of relatively small fractions of nutrients from the forest would promote a remarkable fertilization of the seawater in the Inlet. If this interpretation is correct, inputs of organic substances (e.g. humic acids) probably are also relevant in Bananal Inlet. In this scenario, the forest that surrounds the sampling site could play as an important factor for the input of nutrients into the site. The influence of run-off from an adjacent forest to algal communities has already been shown [43] . These authors demonstrated that the run-off from a forest in the east coast of South Korea promoted a remarkable increase in heavy metals, especially cadmium, detected in the algal flora besides a relevant nutrient enrichment.

A second hypothesis refers to the effects of the water circulation in the region. Despite the Inlet is an inhabited area, it is close to urbanized districts of Maricá, Niterói and Rio de Janeiro municipalities. The short distance to urban areas would favor the input of seawater with high concentrations of nutrients (and even pollutants) in Bananal Inlet. The entrance of Guanabara Bay is ca. 20 km to the Inlet, and the Bay itself is an important source of dissolved nutrients to adjacent areas [44] . These arguments are hypothetic, but there is some evidence to corroborate with this interpretation. For instance, in some field trips it was possible to detect the presence of solid waste (plastic, paper, etc.), in moderate amounts, floating in the Inlet. The occurrence of these records had no apparent link with events such as heavy storms or windy conditions in previous days to the field trip.

Table 3: Some selected mean values for dissolved nutrients collected throughout the present study in the two sampling sites. The results are expressed as μM (except N:P ratio) and represent the average of four replicates $\pm$ SD (n = 4)

Nutrient							
Sampling	N-Ammonia	N-Nitrite	N-Nitrate	N-Urea	Total N	Phosphate	N:P ratio
Dec/2000							
Itapuca Stone	14.7 ± 4.65	0.70 ± 0.14	3.02 ± 1.76	1.85 ± 0.38	20.2 ± 3.47	2.68 ± 0.83	7.56 ± 1.89
Bananal Inlet	4.08 ± 0.15	0.51 ± 0.12	3.16 ± 0.32	0.88 ± 0.19	8.63 ± 0.36	1.16 ± 0.37	7.44 ± 2.39
Mar/2001							
Itapuca Stone	2.80 ± 0.47	0.34 ± 0.06	2.42 ± 0.77	0.89 ± 0.13	6.45 ± 1.80	0.98 ± 0.32	6.58 ± 1.39
Bananal Inlet	1.83 ± 0.21	0.30 ± 0.05	3.96 ± 0.49	0.34 ± 0.08	6.43 ± 0.52	0.56 ± 0.22	11.5 ± 7.79
Jun/2001							
Itapuca Stone	5.31 ± 3.71	1.72 ± 0.15	8.72 ± 3.64	2.31 ± 0.52	18.1 ± 6.76	1.91 ± 0.54	9.48 ± 2.12
Bananal Inlet	0.87 ± 0.13	0.83 ± 0.10	4.61 ± 0.14	1.11 ± 0.20	7.42 ± 0.10	1.14 ± 0.14	6.51 ± 0.71
Nov/2001							
Itapuca Stone	2.72 ± 0.10	1.21 ± 0.06	7.03 ± 0.18	1.37 ± 0.31	12.4 ± 0.15	1.60 ± 0.10	7.73 ± 0.39

Bananal Inlet	2.84 ± 0.27	0.86 ± 0.02	8.90 ± 0.28	1.07 ± 0.28	13.7 ± 0.19	1.59 ± 0.39	8.60 ± 2.10
Mar/2002							
Itapuca Stone	5.74 ± 0.76	1.02 ± 0.08	4.78 ± 0.11	1.15 ± 0.36	19.1 ± 9.2	3.93 ± 0.47	19.1 ± 9.2
Bananal Inlet	1.25 ± 0.14	0.30 ± 0.09	4.58 ± 0.48	0.53 ± 0.03	10.9 ± 0.7	2.38 ± 0.22	20.6 ± 1.1
Jun/2002							
Itapuca Stone	13.5 ± 1.67	3.28 ± 0.28	6.45 ± 3.66	2.30 ± 0.18	32.1 ± 2.7	4.45 ± 0.82	14.0 ± 1.1
Bananal Inlet	1.30 ± 0.16	0.23 ± 0.04	3.97 ± 0.43	0.51 ± 0.06	8.8 ± 1.1	1.66 ± 0.60	17.2 ± 1.4
Aug/2002							
Itapuca Stone	7.11 ± 1.60	1.83 ± 0.38	12.4 ± 4.66	1.71 ± 0.51	24.7 ± 6.74	2.39 ± 0.66	10.3 ± 0.5
Bananal Inlet	0.90 ± 0.25	0.25 ± 0.04	2.25 ± 0.62	1.37 ± 0.58	6.1 ± 1.89	0.48 ± 0.02	12.9 ± 4.2
Oct/2002							
Itapuca Stone	8.60 ± 1.84	2.00 ± 0.07	9.79 ± 2.09	2.13 ± 0.55	24.6 ± 4.48	1.34 ± 0.09	18.3 ± 2.7
Bananal Inlet	0.76 ± 0.15	0.16 ± 0.06	1.96 ± 0.40	1.26 ± 0.16	5.4 ± 0.53	0.77 ± 0.06	7.1 ± 1.1
Jan/2003							
Itapuca Stone	4.91 ± 0.55	0.44 ± 0.07	1.88 ± 0.43	1.87 ± 0.33	11.0 ± 0.76	1.29 ± 0.29	8.7 ± 1.2
Bananal Inlet	1.31 ± 0.29	0.22 ± 0.07	2.47 ± 0.60	2.27 ± 0.25	8.5 ± 0.99	0.56 ± 0.07	15.5 ± 3.5
Mar/2003							
Itapuca Stone	10.5 ± 3.69	0.70 ± 0.34	3.42 ± 0.24	1.29 ± 0.52	20.3 ± 6.23	2.87 ± 1.20	17.0 ± 4.8
Bananal Inlet	3.95 ± 0.83	0.91 ± 0.69	3.59 ± 1.03	0.67 ± 0.16	11.2 ± 1.36	1.39 ± 0.55	17.3 ± 4.3
Jun/2003							
Itapuca Stone	14.0 ± 0.81	1.57 ± 0.15	4.48 ± 0.37	1.01 ± 0.12	22.6 ± 1.53	1.27 ± 0.40	22.6 ± 2.5
Bananal Inlet	1.46 ± 0.79	0.43 ± 0.03	2.10 ± 0.17	0.36 ± 0.08	5.5 ± 2.25	0.75 ± 0.71	15.1 ± 5.2
Aug/2003							
Itapuca Stone	8.96 ± 0.52	1.72 ± 0.07	6.24 ± 0.73	0.79 ± 0.09	18.6 ± 0.51	0.81 ± 0.03	23.6 ± 2.4
Bananal Inlet	1.44 ± 1.17	0.29 ± 0.03	1.93 ± 0.23	0.39 ± 0.08	4.9 ± 1.67	0.63 ± 0.31	13.5 ± 6.7
Nov/2003							

Itapuca Stone	10.6 ± 1.30	3.04 ± 0.73	11.1 ± 4.18	2.41 ± 0.54	28.0 ± 4.5	1.65 ± 0.35	11.8 ± 0.8
Bananal Inlet	2.12 ± 0.54	0.58 ± 0.07	1.91 ± 0.14	0.69 ± 0.23	7.3 ± 1.1	1.34 ± 0.37	11.6 ± 4.7
Feb/2004							
Itapuca Stone	3.73 ± 0.59	0.72 ± 0.15	1.39 ± 0.70	1.36 ± 0.31	18.3 ± 3.25	6.25 ± 1.68	14.3 ± 5.0
Bananal Inlet	1.87 ± 0.80	0.47 ± 0.06	1.89 ± 0.80	0.50 ± 0.16	6.5 ± 0.6	1.15 ± 0.27	14.5 ± 6.1

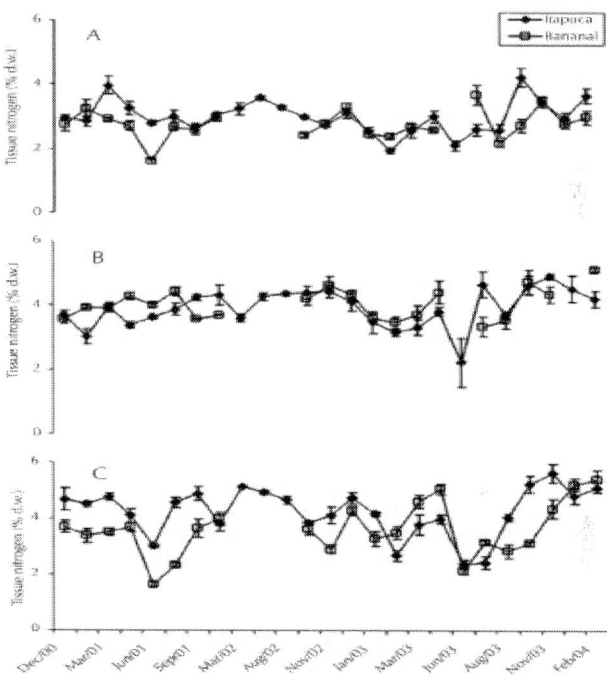

Figure 2: Temporal fluctuations in tissue nitrogen of Chaetomorpha antennina (A), Gymnogongrus griffthisiae (B), and Ulva fasciata (C) sampled in Itapuca Stone and Bananal Inlet from December 2000 to February 2004. Data are expressed as percentage of the dry weight (d.w.) and each point represents the mean of four replicates ± standard deviation (n = 4).

Garbage in the area seems to result from peculiar patterns of circulation in the Inlet, since no local source of pollution exists in the site itself. As one admits the transport of solid garbage from adjacent areas to the Inlet, it is presumable to assume that dissolved nutrients from surrounding eutrophic waters could achieve the Inlet. Nevertheless, it is important to reinforce that in general the seawater in the Inlet is predominantly clean and transparent. Moreover, the Inlet has a remarkable wave action, a factor that contributes for a quick dilution of substances and transport of materials, establishing a presumably low residence time in the Inlet.

A third hypothesis is the occurrence of some upwelling events in coastal areas of Niterói municipality. These events frequently reach Bananal Inlet in summer, but rarely could reach Itapuca Stone (located inside Guanabara Bay). For instance, in one of the filed trips (January 9th, 2004) waters of 17°C reached the Inlet (Table 2), a typical temperature of upwelling events in the region. This interpretation is reinforced by the detection of high concentrations of nitrate in that month in the Inlet (3.39 ± 0.83 μM), which were not statistically different of those detected in Itapuca Stone.

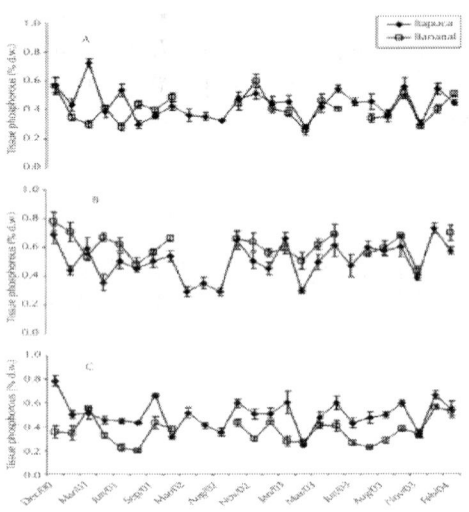

Figure 3: Temporal fluctuations in tissue phosphorus of C. antennina (A), G. griffthisiae (B), and U. fasciata (C) sampled in Itapuca Stone and Bananal Inlet from December 2000 to February 2004. Data are expressed as percentage of the d.w. and each point represents the mean of four replicates ± SD (n = 4).

The excess of nutrients in Guanabara Bay characterizes that environment as eutrophic [30] , achieving hypereutrophy in some parts and generating relevant floristic changes. A small number of macroalgal species exists near the entrance of Guanabara Bay, where Itapuca Stone is located. According to Taouil & Yoneshigue [45] , there are only 45 species in that area, while more than 70 species were recorded in the same site by the end of the decade of 1960. This number contrasts with the 92 species found by Moreira (unpublished data) in Bananal Inlet. The ongoing process of eutrophication has been promoting a loss of biodiversity in Guanabara Bay, changing the characteristics of local algal communities [45]. Opportunistic species, which tolerate high concentrations of pollutants (generally present in large volumes in environments disturbed by cultural eutrophication) tend to proliferate, occupying the space left by more sensitive species [46] Despite significant differences in the concentrations of dissolved nutrients have been detected in the sites, N:P ratio tended to be similar at the sampling sites throughout the study. An overall mean value of 14.9:1 was calculated for Itapuca Stone and 14.7:1 for Bananal Inlet. Compared to the classical studies [47] [48] , which indicate a N:P ratio of 16:1 as an average value for world oceans, the current results are within fluctuations expected for field data. Despite the small number of samples analyzed, values around 15:1 would not indicate limitation by N or P to the algae.

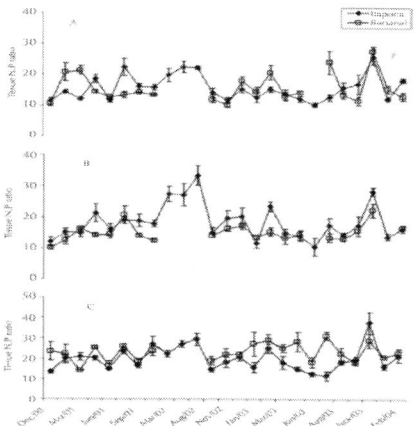

Figure 4: Temporal fluctuations in tissue N:P ratio of C. antennina (A), G. griffthisiae (B), and U. fasciata (C) sampled in Itapuca Stone and Bananal Inlet from December 2000 to February 2004. Data are expressed as percentage of the d.w. and each point represents the mean of four replicates ± SD (n = 4).

This trend contrasts with other Brazilian studies. For instance, Aidar et al. [49] obtained an average N:P ratio of 12:1 for the continental shelf of Ubatuba, São Paulo State, suggesting a phytoplankton limitation by nitrogen. Valentin et al. [30] found wide variations in atomic ratios in different sampling sites in inner parts of Guanabara Bay with a remarkable influence of the tidal regime. Low N:P rations in inner parts of Guanabara Bay (<10:1) were interpreted as a result of excess phosphate from domestic effluents [30] .

Measurements of N:P ratio are insufficient to determine the presence or absence of a given species in an environment under strong impact. It is widely known that each species has an optimal N:P ratio for its metabolic demands [7] [50] , but it is unlike to happen competitive exclusion due to this factor. The exclusion of a given species from a disturbed environment by anthropic action is more likely to be a consequence of the effects of pollutants, but normally it is very difficult to determine the limits for the action of a specific pollutant since in general complex mixtures are discharged into the sea [51] .

Total Tissue Nitrogen and Phosphorus in the Seaweeds

In this study, the red alga G. griffithsiae tended to show higher concentrations of tissue nitrogen and phosphorus than the green algae.

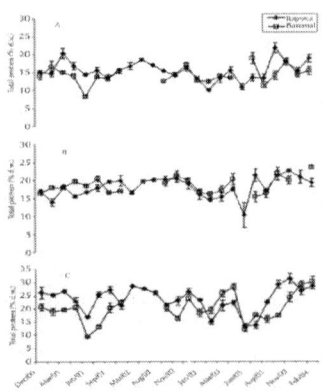

Figure 5: Temporal fluctuations in total protein content of C. antennina (A), G. griffthisiae (B), and U. fasciata (C) sampled in Itapuca Stone and Bananal

Inlet from December 2000 to February 2004. Data are expressed as percentage of the d.w. and each point represents the mean of four replicates ± SD (n = 4).

This is in accordance with studies of Diniz et al. [52] and Lourenço et al. [34] who characterized the chemical composition of seaweeds from Brazilian coastal environments. Rhodophytes tended to show more nitrogen-bearing pigments and higher concentrations of hydrosoluble protein. Higher concentrations of phosphorus in seaweeds would be related to the characteristics of fast growing species, which produce more ATP [7] [52] .

In physiological terms, seaweeds from tropical environments show a low demand for dissolved nutrients, compared to seaweeds from temperate environments [53] [54] . In the tropics plants are commonly saturated with nutrients even in low concentrations (e.g.: 3.0 μM for N e 0.25 μM for P), which are sufficient to generate high growth rates and tissue nutrients in suitable concentrations. Compared to phytoplankton, seaweeds have a high procurement for carbon, higher than the relative demand for nitrogen and phosphorus, a characteristic related to the life cycles, life span, growth and composition of the thalli [55] . If the availability of inorganic nutrients increases temporarily, there is a natural trend of a fast up take and assimilation of nutrients, displaying in the thalli higher concentrations of N and P [56] .

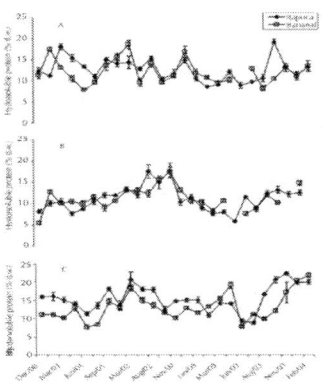

Figure 6: Temporal fluctuations in hydrosoluble protein of C. antennina (A), G. griffthisiae (B), and U. fasciata (C) sampled in Itapuca Stone and Bananal Inlet from December 2000 to February 2004. Data are expressed as percentage of the d.w. and each point represents the mean of four replicates ± SD (n = 4).

However, if an abundant amount is kept for a longer period, there is a trend of saturation of the thalli with nutrients, and no increment in algal responses to nutrients is recorded [53]. Thus, an excess of nutrient in water not necessarily will generate high concentrations of N and P in tissues, because even a luxuriant consumption of nutrients (such as nitrogen) has a limit, without a progressively linear response to the stimulus after a given point. If high concentrations of nutrients persist, the algae (macroand micro-algae) may either excrete inorganic nutrients or keep the synthesis of organic matter in stable levels, without increases in concentrations of Nand P-bearing-substances [7] [12] [57] . These are typical responses of algae in eutrophic environments.

In Bananal Inlet, concentrations of dissolved nutrients are supposedly enough to sustain an optimum growth in the species tested. The observations of the local species showed tissue-N concentrations never lower than 2% d.w., suggesting that growth conditions would be suitable throughout the year [58] . Thus, the seaweeds tend to show high tissue N and P concentrations, possibly close to the saturation level. In this context, measurements done with samples from Bananal Inlet tended to be predominantly high, similar to those of Itapuca Stone.

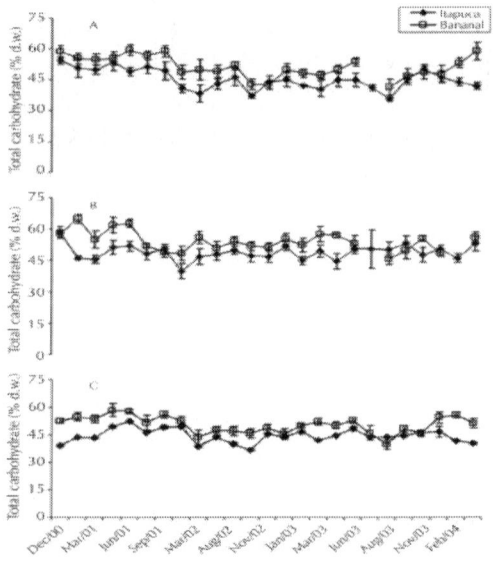

Figure 7: Temporal fluctuations in total carbohydrate of C. antennina (A), G. griffthisiae (B), and U. fasciata (C) sampled in Itapuca Stone and Bananal Inlet

from December 2000 to February 2004. Data are expressed as percentage of the d.w. and each point represents the mean of four replicates ± SD (n = 4).

Another factor to contribute to diminish differences of tissue N and P measured in the different sites are topographical features. Itapuca Stone is plan, with low natural shelters (e.g. crevices in rocks), and seaweeds are directly exposed to dryness during low tides. This condition promotes a strong stress in the species, which is expressed as damages in the thalli and loss of tissue nutrients [59] [60] . In many field trips in summer months and also in short isolated periods of strong heat in any season, several individuals showed tips bleached, indicating loss of their constituents. This phenomenon was particularly common in G. griffithsiae, especially easy to see due the contrast between the dark red of the healthy individuals and the pale color (white or yellowish) of damaged plants. In Bananal Inlet this phenomenon was less common, although in some occasions algae were found with bleaching especially after period of strong heat. There, U. fasciata was the species that showed more commonly damaged thalli, while G. griffithsiae has never been found with bleaching in Bananal Inlet. This trend is possibly a consequence of the specific occupation of the space by the red alga in Bananal Inlet, always in specific places sheltered from sunshine, under shadows created by large rocks and crevices. These arguments are important to understand why concentrations of tissue N and P of G. griffithsiae were similar in both environments (Figure 2 and Figure 3) and how abiotic sources of stress may influence the nutrient composition of the algae.

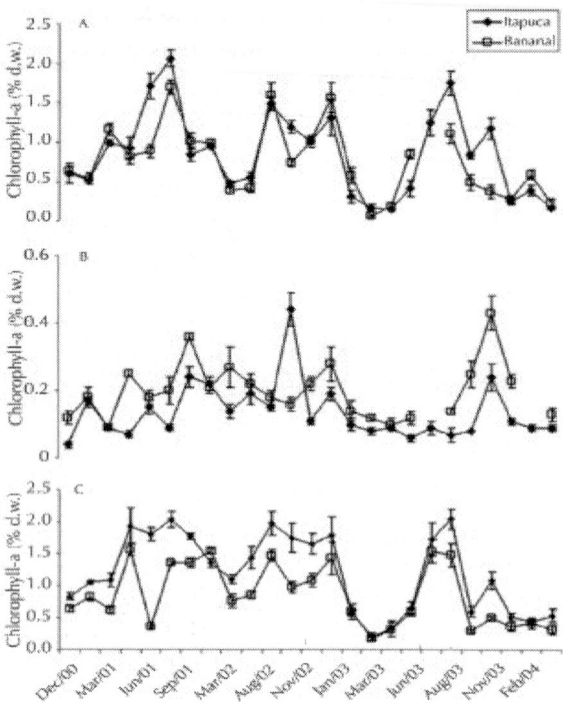

Figure 8: Temporal fluctuations in chlorophyll a content of C. antennina (A), G. griffthisiae (B), and U. fasciata (C) sampled in Itapuca Stone and Bananal Inlet from December 2000 to February 2004. Data are expressed as percentage of the d.w. and each point represents the mean of four replicates ± SD (n = 4).

Despite the minor differences in tissue N and P for each species in both sites, most of the observations did not reveal cyclic patterns of variation of concentrations, except for lower values of chlorophyll a and total carotenoid in summer/early autumn, contrasting with typical results reported for temperate environments [5] [8] . Fluctuations found in tropical environments are associated to significant changes in concentrations of nutrients throughout the year (including possible inputs of nutrients into the system) or the occurrence of a more intense environmental factor (eg.: temperature, upwelling), affecting algal responses during part of the annual cycle. In a related study, Lourenço et al. [15] studied the seasonal variations of tissue N and P of eight macroalgal species of Araruama Lagoon, a hypersaline environment in Rio de Janeiro State. Remarkable seasonal variations in tissue nutrients for the seaweeds were found, with higher values in autumn and lower in spring for most of the species. The authors also considered that seaweeds are severely affected by high temperatures, at least in part of the spring and in the summer. The absence of patterns for seasonal variations in tissue N and P of the seaweeds in the present study suggests: (i) that the nutrient supply is virtually constant or it suffers minor variations; and (ii) that other abiotic factors (e.g. temperature) play a secondary role to influence nutrient accumulation by the seaweeds. The lack of seasonal variations of tissue N and P of 10 seaweeds (6 green and 4 red algae) was also confirmed [56] in a seven-year study (from 1997 to 2004) performed in Boa Viagem Beach, a site located in Guanabara Bay.

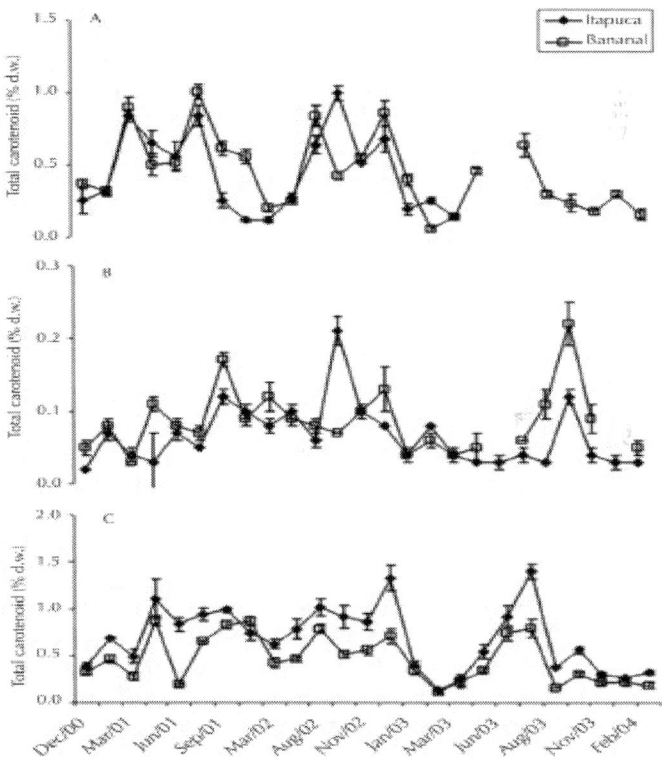

Figure 9: Temporal fluctuations in total carotenoid content of C. antennina (A), G. griffthisiae (B), and U. fasciata (C) sampled in Itapuca Stone and Bananal Inlet from December 2000 to February 2004. Data are expressed as percentage of the d.w. and each point represents the mean of four replicates ± SD (n = 4).

Lourenço et al. [28] found N:P atomic ratios in the algal tissues typically higher than 20:1 and lower phosphorus concentrations in the water than at Itapuca Stone in the present study.

According to the Björnsäter and Wheeler's classification [61] of macroalgal nutrient status based on N:P ratio of tissues, a N:P ratio < 16 indicates N-limitation; a N:P ratio 16 - 24 indicates N-sufficiency and P-sufficiency —i.e. no limitation and N:P > 24 indicates P-limitation. Applying this classification to our data we could conclude that the macroalgae in the sampling sites are permanently N and P-sufficient, with few exceptions. However, the N:P ratio must be evaluated with care, as it may obscure trends for the individual elements. For instance, the lower values for phosphorus in the seaweeds were normally > 0.40% d.w. A 0.40% of tissue P does not represent a low level of phosphorus, and it is actually higher than values found for many other algae from tropical environments [53] [54] . In some cases a high N:P ratio observed may be strongly affected by the high concentrations of nitrogen and is not necessarily indicative of P limitation. Thus, the classification of Björnsäter & Wheeler [61] must be considered with caution, because the ranges may not be suitable for macroalgae from tropical environments such as Guanabara Bay and Bananal Inlet. Further investigations are needed to test the suitability of that classification for tropical environments, where seaweeds typically grow well with low concentrations of dissolved nutrients and normally have lower tissue N and P compared to species from temperate environments.

Protein, Carbohydrate and Photosynthetic Pigments

Following the same general trends described for total nitrogen, the red alga G. griffithsiae tended to show higher concentrations of hydrosoluble protein than the green algae. Our results also agree with those of Gressler et al. [62] , who found that four red seaweeds from Brazil typically show hydrosoluble protein fluctuating from 4.6% to 18.3% of d.w. In our study, hydrosoluble protein of G. griffithsiae fluctuates from 10% to 20% of d.w in most of the observations.

Possibly most of the studies on major chemical components of seaweeds have focus on the nutritional properties of the species, e.g. [63] -[69] . However, analyses of major chemical components are important tools for environmental issues. Previous studies performed confirm that tissue protein is positively correlated dissolved nitrogen in the water [70] [71] . In the present study, the apparent permanent sufficiency on nutrients (especially nitrogen) would contribute to the high measurements of protein in the seaweeds in both sites. One can speculate that in Itapuca Stone the saturating levels of dissolved nutrients would keep protein in high concentrations. Despite the concentration of nutrients is not so high at Bananal Inlet it would be enough to generate a high accumulation of protein. These interpretations have support from the studies of [53] with the green alga Enteromorpha intestinalis (=Ulva intestinalis) in mesocosms, in which the alga did not respond to enrichment with nutrients if thalli concentrations were saturated.

The accumulation of protein tends to promote a decrease in carbohydrate production. The assimilation of nitrogen (ammonia) into amino acids occurs via a GS/GOGAT (glutamine sintetase/glutamine: 2-oxoglutarate aminotransferase) system, resulting in production of glutamate. For the synthesis of glutamate two molecules of 2-oxiglutarate are required, while for the synthesis of other amino acids carbon skeletons are required through respiratory chain. As a result, assimilation photosynthetic nitrogen stimulates the respiratory flux of carbon. In cells growing with high concentrations of nutrients, levels of endogenous reserves of carbohydrate drop and the assimilation of nitrogen in amino acids depends on recent photosynthesis [10] [72] .

These arguments support the occurrence of higher concentrations of carbohydrate in samples from Bananal Inlet, especially G. griffithsiae and U. fasciata. Results for G. griffithsiae are similar to those of Perfeto [73] , who found values predominantly >50% d.w. for the same species in a seasonal study in southern Brazil, under subtropical climate. Pádua et al. [24] also reported similar results for total carbohydrate, with concentrations varying from 55.3% to 58.4% of d.w. for Ulva lactuca and U. fasciata from Paraná State, Brazil. Protein levels measured in those species by the same authors varied from 13.3% to 18.4% d.w. [24] ; these values are slightly lower than the current results. Higher concentrations of carbohydrate than that of protein in 30 common seaweeds of tropical Australia were also found [74] , as well as for three common species of Abu Qir Bay, Egipt [75] . Despite the amount of dissolved nutrients in Bananal Inlet is relatively high, it is lower than in Itapuca Stone. Considering the coupling between carbon and nitrogen metabolism, it is logical to understand a tendency for more carbohydrate in Bananal Inlet, even with discrete differences in some comparisons. As C. antennina exhibited the smallest differences for virtually all comparisons of the two sites, this alga probably has a naturally low demand for nutrients. Supposedly the chemical composition of C. antennina was virtually not affected by differences in nutrient regimes in the sites. This trend has been documented for slow-growing tropical macroalgae, such as thus of the genus Sargassum [54] and green algae typical of warm waters, such as Halimeda [53] . Thus, independent of specific environmental characteristics in which they are, these species tend to exhibit discrete responses of synthesis of substances to available nutrients, keeping their chemical composition with slight fluctuations.

A wide range of variation in the content of hydrosoluble protein in the green algae U. fasciata and C. antennina agrees with the variations in tissue nitrogen throughout time, with accumulation of nitrogen in some periods as protein. The occurrence of very "flexible" protein content in those seaweeds points to the capability of them to respond to rapid environmental changes. Fleurence [76] points that protein contents in Ulva typically vary from 10% to 26% d.w.

The high concentrations of total carotenoid found in this study (normally higher than 50% of the chlorophyll content) points to the role of carotenoids as shields to protect the photosystems [77] . The presence of different quantities and kinds of pigments (chlorophyll, carotenoids, phycoerythrin) in G. griffithsiae results in a high capacity to absorb light in virtually all visible light. The diverse pigments of the red alga may favor the species to occupy microhabitats not directly exposed to light in rocky shores, under shadows of large rocks or in crevices. G. griffithsiae is found in these microhabitats at Bananal Inlet. In this context, the species could have competitive advantages for not exposing itself to high intensities of light. Presumably, G. griffithsiae shows an efficient apparatus for light absorption. The presence of accessory pigments could account for the lower concentrations of chlorophyll in G. griffithsiae compared to the other species.

Remarkable oscillations in pigment content were recorded in Ulva fasciata and they seem to be related to the loss of pigments in certain periods of the year, as a consequence of partial loss of thalli due to excessive desiccation. Intertidal seaweeds experience extreme conditions of heat in tropical environments, which may affect their morphological features [78] . As a foliose alga that occupies the mid-littoral area, U. fasciata is particularly exposed to high temperatures. This factor is apparently less important in C. antennina, which inhabits places under direct wave action and permanently in contact with seawater in movement. Moreover, arguments relative to life cycle of the species (not assessed in this study) are also potentially relevant, especially for U. fasciata, which suffers a population decline in summer due to phenological processes in the region [51] . The apparent biomass fluctuations observed for U. fasciata (with lower biomass in summer) were similar in both environments, suggests that abiotic factors such as light and temperature might be as important as dissolved nutrients to affect the chemical composition of the species, as demonstrated for Gracilaria tikvahiae [59] . Gymnogongrus griffithsiae also presented significant changes in the concentrations of photosynthetic pigments (less than U. fasciata) and loss of thalli in samples collected at Itapuca Stone, after periods of strong heat. The exposure of seaweeds to high irradiation in summer could account for the lower contents of chlorophyll recorded for G. griffithsiae and U. fasciata due to loss of tissues. Aguilera et al. [14] recorded this same trend for Porphyra umbilicalis from the North Sea, with loss of chlorophyll after periods of intense heat.

CONCLUDING REMARKS

Changes in the concentrations of total protein, hydrosoluble protein, total carbohydrate, chlorophyll a, total carotenoid, tissue nitrogen and tissue phosphorus in Chaetomorpha antennina, Gymnogongrus griffithsiae and Ulva fasciata were predominantly small or absent in the two sampling sites. No clear cyclic variations throughout time were detected for the substances measured in the seaweeds, except for pigments, which showed declines at the end of summer months. Dissolved nutrients are available in higher concentrations to seaweeds at Itapuca Stone, where they possibly achieve permanent saturating levels for the seaweeds. Concentrations of dissolved nitrogen and phosphorus at Bananal Inlet seem to be always high enough to supply the metabolic demands of the seaweeds for synthesis of organic substances and growth, with no evidence of nutrient limitation throughout the year.

ACKNOWLEDGEMENTS

We are indebted to FAPERJ (Rio de Janeiro State Research Support Foundation) for the financial support to this study. We acknowledge Dr Aguinaldo N. Marques Júnior for offering us laboratory facilities. The authors are grateful to Leonardo de Souza (deceased), Amanda P. de Freitas, Frederico A. Herdy, Rodrigo R. Machado, Giselle S. Costard, Rachel S. Guanabara, Marcello T. Carvalho, and Joana N.P. Freitas their assistance in field trips and in laboratory analyses. We thank Dr. Maria Teresa M. de Széchy, Dr. Ricardo Coutinho and Dr. Roberto C. Villaça for their critical comments on the early draft of this study (in Portuguese).

REFERENCES

1. Lapointe, B.E. and Duke, S.C. (1984) Biochemical Strategies for Growth of Gracilaria tikvahiae (Rhodophyta) in Relation to Light Intensity and Nitrogen Availability. Journal of Phycology, 20, 488-495. http://dx.doi.org/10.1111/j.0022-3646.1984.00488.x

2. Valiela, I., McClelland, J., Hauxwell, J., Behr, P.J. and Foreman, K. (1997) Macroalgal Blooms in Shallow Estuaries: Controls and Ecophysiological and Ecosystem Consequences. Limnology and Oceanography, 42, 1105-1118.http://dx.doi.org/10.4319/lo.1997.42.5_part_2.1105

3. Rivers, J.S. and Peckol, P. (1995) Interactive Effects of Nitrogen and Dissolved Inorganic Carbon on Photosynthesis, Growth, and Ammonium Uptake of the Macroalgae Cladophora vagabunda and Gracilaria tikvahiae. Marine Biology, 121, 747-753.http://dx.doi.org/10.1007/BF00349311

4. Fong, P., Donohoe, R.M. and Zedler, J.B. (1994) Nutrient Concentration in Tissue of the Macroalga Enteromorpha as a Function of Nutrient History: An Experimental Evaluation Using Field Microcosms. Marine Ecology Progress Series, 106, 273-281.http://dx.doi.org/10.3354/meps106273

5. Wheeler, P.A. and Björnsäter, B.R. (1992) Seasonal Fluctuations in Tissue Nitrogen, Phosphorus, and N:P for Five Macroalgal Species

Common to the Pacific Northwest Coast. Journal of Phycology, 28, 1-6. http://dx.doi.org/10.1111/j.0022-3646.1992.00001.x

6. Villares, R. and Carballeira, A. (2003) Seasonal Variation in the Concentrations of Nutrients in Two Green Macroalgae and Nutrient Levels in Sediments in the Rias Baixas (NW Spain). Estuarine, Coastal and Shelf Science, 58, 887-900.http://dx.doi.org/10.1016/j.ecss.2003.07.004

7. Lobban, C.S. and Harrison, P.H. (1994) Seaweed Ecology and Physiology. Cambridge University Press, New York. http://dx.doi.org/10.1017/CBO9780511626210

8. Peckol, P., DeMeo-Andersen, B., Rivers, J., Valiela, I., Maldonado, M. and Yates, J. (1994) Growth, Nutrient Uptake Capacities and Tissue Constituents of the Macroalgae Cladophora vagabunda and Gracilaria tikvahiae Related to Site-Specific Nitrogen Loading Rates. Marine Biology, 121, 175-185. http://dx.doi.org/10.1007/BF00349487

9. Kamer, K., Fong, P., Kennison, R. and Schiff, K. (2004) Nutrient Limitation of the Macroalga Enteromorpha intestinalis Collected along a Resource Gradient in a Highly Eutrophic Estuary. Estuaries, 27, 201-208. http://dx.doi.org/10.1007/BF02803377

10. Turpin, D.H. (1991) Effects of Inorganic N Availability on Algal Photosynthesis and Carbon Metabolism. Journal of Phycology, 27, 14-20. http://dx.doi.org/10.1111/j.0022-3646.1991.00014.x

11. Bird, K.T., Habig, C. and DeBusk, T. (1982) Nitrogen Allocation and Storage Patterns in Gracilaria tikvahiae (Rhodophyta). Journal of Phycology, 18, 344-348.http://dx.doi.org/10.1111/j.1529-8817.1982.tb03194.x

12. Lourenço, S.O., Barbarino, E., Lavín, P.L., Marquez, U.M.L. and Aidar, E. (2004) Distribution of Intracellular Nitrogen in Marine Microalgae: Calculation of New Nitrogen-to-Protein Conversion Factors. European Journal Phycology, 39, 17-32.http://dx.doi.org/10.1080/0967026032000157156

13. Dere, S., Dalkiran, N., Karacaoğlu, D., Yildiz, G. and Dere, E. (2003) The Determination of Total Protein, Total Soluble Carbohydrate and Pigment Contents of Some Gemlik-Karacaali (Bursa) and Erdek-Ormanli (Balikesir) in the Sea Marmara, Turkey. Oceanologia, 45, 453-471.

14.   Aguilera, J., Jiménez, C., Figueroa, F.L., Lebert, M. and Häder, D.P. (1999) Effect of Ultraviolet Radiation on Thallus Absorption and Photosynthetic Pigments in the Red Alga Porphyra umbilicalis. Journal of Photochemistry and Photobiology B: Biology, 48, 75-82.http://dx.doi.org/10.1016/S1011-1344(99)00015-9

15.   Lourenço, S.O., Barbarino, E., Nascimento, A. and Paranhos, R. (2005) Seasonal Variations in Tissue Nitrogen and Phosphorus of Eight Macroalgae from a Tropical Hypersaline Coastal Environment. Cryptogamie Algologie, 26, 355-371.

16.   Aitken, K.A., Melton, L.D. and Brown, M.T. (1991) Seasonal Protein Variation in the New Zealand Seaweeds Porphyra columbina Mont. and Porphyra subtumens J. Ag. (Rhodophyceae). Japanese Journal of Phycology, 39, 307-317.

17.   Henley, W.J. and Dunton, K.H. (1995) A Seasonal Comparison of Carbon, Nitrogen, and Pigment Content in Laminaria solidungula and L. saccharina (Phaeophyta) in the Alaskan Artic. Journal of Phycology, 31, 325-331. http://dx.doi.org/10.1111/j.0022-3646.1995.00325.x

18.   Korbee, N., Figueroa, F.L. and Aguilera, J. (2005) Effect of Light Quality on the Accumulation of Photosynthetic Pigments, Proteins and Mycosporine-Like Amino Acids in the Red Alga Porphyra leucosticta (Bangiales, Rhodophyta). Journal of Photochemistry and Photobiology B: Biology, 80, 71-78.http://dx.doi.org/10.1016/j.jphotobiol.2005.03.002

19.   Tabarsa, M., Rezaei, M., Ramezanpour, Z., Waaland, J.R. and Rabiei, R. (2012) Fatty Acids, Amino Acids, Mineral Contents, and Proximate Composition of Some Brown Seaweeds. Journal of Phycology, 48, 285-292. http://dx.doi.org/10.1111/j.1529-8817.2012.01122.x

20.   Madden, M., Mitra, M., Ruby, D. and Schwarz, J. (2012) Seasonality of Selected Nutritional Constituents of Edible Delmarva Seaweeds. Journal of Phycology, 48, 1289-1298. http://dx.doi.org/10.1111/j.1529-8817.2012.01207.x

21.   Polat, S. and Ozogul, Y. (2013) Seasonal Proximate and Fatty Acid Variations of Some Seaweeds from the Northeastern Mediterranean Coast. Oceanologia, 55, 375-391.http://dx.doi.org/10.5697/oc.55-2.375

22. Schaffelke, B. (1999) Short-Term Nutrient Pulses as Tools to Assess Responses of Coral Reef Macroalgae to Enhanced Nutrient Availability. Marine Ecology Progress Series, 182, 305-310. http://dx.doi.org/10.3354/meps182305

23. Wong, K.H. and Cheung, C.K. (2001) Nutritional Evaluation of Some Subtropical Red and Green Seaweeds Part II. In Vitro Protein Digestibility and Amino Acid Profiles of Protein Concentrates. Food Chemistry, 72, 11-17. http://dx.doi.org/10.1016/S0308-8146(00)00176-X

24. Pádua, M., Fontoura, P.S.G. and Mathias, A.L. (2004) Chemical Composition of Ulvaria oxysperma (Kützing) Bliding, Ulva lactuca (Linnaeus) and Ulva fasciata (Delile). Brazilian Archives of Biology and Technology, 47, 49-55. http://dx.doi.org/10.1590/S1516-89132004000100007

25. Devi, G.K., Thirumaran, G., Manivannan, K. and Anantharaman, P. (2009) Element Composition of Certain Seaweeds from Gulf of Mannar Marine Biosphere Reserve; Southeast Coast of India. World Journal of Dairy & Food Sciences, 4, 46-55.

26. Diniz, G.S., Barbarino, E., Oiano-Neto, J., Pacheco, S. and Lourenço, S.O. (2011) Gross Chemical Profile and Calculation of Nitrogen-to-Protein Conversion Factors for Five Tropical Seaweeds. American Journal of Plant Sciences, 2, 287- 296. http://dx.doi.org/10.4236/ajps.2011.23032

27. Siddique, M.A.M., Aktar, M. and Khatib, M.A.M. (2013) Proximate Chemical Composition and Amino Acid Profile of Two Red Seaweeds (Hypnea pannosa and Hypnea musciformis) Collected from St. Martin's Island, Bangladesh. Journal of Fisheries Sciences, 7, 178-186.

28. Lourenço, S.O., Barbarino, E., Nascimento, A., Freitas, J. and Diniz, G. (2006) Tissue Nitrogen and Phosphorus in Seaweeds in a Tropical Eutrophic Environment: What a Long-Term Study Tells Us. Journal of Applied Phycology, 18, 389-398. http://dx.doi.org/10.1007/s10811-006-9035-9

29. Mayr, L.M., Tenenbaum, D.R., Villac, M.C., Paranhos, R., Nogueira, C.R., Bonecker, S.L.C. and Bonecker, A.C. (1989) Hydrological Characterization of Guanabara Bay. In: Magoo,

O.T. and Neves, C., Eds., Coastlines of Brazil, American Society of Civil Engineering, New York, 129-139.

30. Valentin, J.L., Tenenbaum, D.R., Bonecker, A.C.T., Bonecker, S.L.C., Nogueira, C.R. and Villac, M.C. (1999) O Sistema Planctônico da Baía de Guanabara: Síntese do Conhecimento. Oecologia Brasiliensis, 7, 35-59.http://dx.doi.org/10.4257/oeco.1999.0701.02

31. Wynne, M.J. (1998) A Checklist of Benthic Marine Algae of the Tropical and Subtropical Western Atlantic: First Revision. In: Nova Hedwigia, Suppl. 116, J. Cramer in der Gebr. Borntraeger Verlagsbuchhandlung, Stuttgart, 1-155.

32. Hach, C.C., Bowden, B.K., Kopelove, A.B. and Brayton, S.T. (1987) More Powerful Peroxide Kjeldahl Digestion Method. Journal of the Association of Official Analytical Chemistry, 70, 783-787.

33. Lowry, O.H., Rosebrough, N.J., Farr, A.L. and Randall, R.L. (1951) Protein Measurement with the Folin Phenol Reagent. The Journal of Biological Chemistry, 193, 265-275.

34. Lourenço, S.O., Barbarino, E., De-Paula, J.C., Da S. Pereira, L.O. and Lanfer Marquez, U.M. (2002) Amino Acid Composition, Protein Content, and Calculation of Nitrogen-to-Protein Conversion Factors for Nineteen Tropical Seaweeds. Phycological Research, 50, 233-241. http://dx.doi.org/10.1111/j.1440-1835.2002.tb00156.x

35. Myklestad, S. and Haug, A. (1972) Production of Carbohydrates by the Marine Diatom Chaetoceros affinis var. willie (Gran) Hustedt. I. Effect of the Concentration of Nutrients in the Culture Medium. Journal of Experimental of Marine Biology and Ecology, 9, 125-136.http://dx.doi.org/10.1016/0022-0981(72)90041-X

36. DuBois, M., Gilles, K.A., Hamilton, J.K., Rebers, P.A. and Smith, F. (1956) Colorimetric Method for Determination of Sugars and Related Substances. Analytical Chemistry, 28, 350-356. http://dx.doi.org/10.1021/ac60111a017

37. Lorenzen, C.J. (1967) Determination of Chlorophyll and Pheopigments: Spectrophotometric Equations. Limnology and Oceanography, 12, 343-346.http://dx.doi.org/10.4319/lo.1967.12.2.0343

38. Strickland, J.D.H. and Parsons, T.R. (1968) A Practical Handbook of Seawater Analysis. Bulletin of Fisheries Research Board of Canada, 167, 1-311.

39. Parsons, T.R., Maita, Y. and Lalli, C.M. (1984) A Manual of Chemical and Biological Methods for Seawater Analysis. Pergamon Press, Oxford.

40. Aminot, A. and Chaussepied, M. (1983) Manuel des Analyses Chimiques en Milieu Marin. CNEXO, Brest.

41. Grasshoff, K., Ehrhardt, M. and Kremling, K. (1983) Methods of Seawater Analysis. Verlag Chemie, Weinheim.

42. Zar, J.H. (1996) Biostatistical Analysis. 3rd Edition, Prentice Hall, Inc., Upper Saddle River.

43. Shin, H.W., Sidharthan, M. and Young, K.S. (2002) Forest Fire Ash Impact on Microand Macroalgae in the Receiving Waters of the East Coast of South Korea. Marine Pollution Bulletin, 45, 203-209. http://dx.doi.org/10.1016/S0025-326X(02)00156-X

44. Kjerfve, B., Ribeiro, C.H.A., Dias, G.T.M., Filippo, A.M. and Quaresma, V.S. (1997) Oceanographic Characteristics of an Impacted Coastal Bay: Baía de Guanabara, Rio de Janeiro, Brazil. Continental Shelf Research, 17, 1609-1643. http://dx.doi.org/10.1016/S0278-4343(97)00028-9

45. Taouil, A. and Yoneshigue-Valentin, Y. (2002) Alterações na Composição Florística das Algas da Praia de Boa Viagem (Niterói, RJ). Revista Brasileira de Botânica, 25, 405-412. http://dx.doi.org/10.1590/S0100-84042002012000004

46. Martínez-Aragón, J.F., Hernández, I., Pérez-Lloréns, J.L., Vázquez, R. and Vergara, J.J. (2002) Biofiltering Efficiency in Removal of Dissolved Nutrients by Three Species of Estuarine Macroalgae Cultivated with Sea Bass (Dicentrarchus labrax) Waste Waters 1.Phosphate. Journal of Applied Phycology, 14, 365-374. http://dx.doi.org/10.1023/A:1022134701273

47. Redfield, A.C. (1958) The Biological Control of Chemical Factors in the Environment. American Scientist, 46, 205- 221.

48. Redfield, A.C., Ketchum, B.H. and Richards, F.A. (1963) The Influence of Organisms on the Chemical Composition of Seawater. In: Hill, M.N., Ed., The Sea, Interscience, New York, 26-77.

49.   Aidar, E., Gaeta, S.A., Giansella-Galvão, S., Kutner, M.B.B. and Teixeira, C. (1993) Ecossistema Costeiro Subtropical: Nutrientes Dissolvidos, Fitoplâncton e Clorofila-a e suas Relações com as Condições Oceanográficas na Região de Ubatuba, SP. Instituto Oceanográfico da USP, 10, 9-43.

50.   Fong, P., Zedler, J.B. and Donohoe, R.M. (1993) Nitrogen vs. Phosphorus Limitation of Algal Biomass in Shallow Coastal Lagoons. Limnology and Oceanography, 38, 906-923.http://dx.doi.org/10.4319/lo.1993.38.5.0906

51.   Teixeira, V.L., Pereira, R.C., Marques Jr., A.N., Leitão Filho, C.M. and Silva, C.A.R. (1987) Seasonal Variations in Infralitoral Seaweed Communities under a Pollution Gradient in Baía de Guanabara, Rio de Janeiro (Brazil). Ciência e Cultura, 39, 423-428.

52.   Diniz, G.S., Barbarino, E. and Lourenço, S.O. (2012) On the Chemical Profile of Marine Organisms from Coastal Subtropical Environments: Gross Composition and Nitrogen-to-Protein Conversion Factors. In: Marcelli, M., Ed., Oceanography, InTech, Rijeka, 297-320. http://dx.doi.org/10.5772/29294

53.   Fong, P., Boyer, K.E., Kamer, K. and Boyle, K.A. (2003) Influence of Initial Tissue Nutrient Status of Tropical Marine Algae on Response to Nitrogen and Phosphorus Additions. Marine Ecology Progress Series, 262, 111-123.http://dx.doi.org/10.3354/meps262111

54.   Hwang, R.L., Tsai, C.C. and Lee, T.M. (2004) Assessment of Temperature and Nutrient Limitation on Seasonal Dynamics among Species of Sargassum from a Coral Reef in Southern Taiwan. Journal of Phycology, 40, 463-473. http://dx.doi.org/10.1111/j.1529-8817.2004.03086.x

55.   Pedersen, M.F. and Borum, J. (1996) Nutrient Control of Algal Growth in Estuarine Waters. Nutrient Limitation and the Importance of Nitrogen Requirements and Nitrogen Storage among Phytoplankton and Species of Macroalgae. Marine Ecology Progress Series, 142, 261-272. http://dx.doi.org/10.3354/meps142261

56.   Kamer, K. and Fong, P. (2001) Nutrient Enrichment Ameliorates the Negative Effects of Reduced Salinity on the Green Macroalga Enteromorpha intestinalis. Marine Ecology Progress Series, 218, 87-93. http://dx.doi.org/10.3354/meps218087

57. Gordon, D.M., Birch, P.B. and McComb, A.J. (1981) Effects of Inorganic Phosphorus and Nitrogen on the Growth of an Estuarine Cladophora in Culture. Botanica Marina, 24, 93-106. http://dx.doi.org/10.1515/botm.1981.24.2.93

58. Hanisak, M.D. (1979) Nitrogen Limitation of Codium fragile ssp. tomentosoides as Determined by Tissue Analysis. Marine Biology, 50, 333-337. http://dx.doi.org/10.1007/BF00387010

59. Hanisak, M.D. (1993) Nitrogen Release from Decomposing Seaweeds: Species and Temperature Effects. Journal of Applied Phycology, 5, 175-181. http://dx.doi.org/10.1007/BF00004014

60. Menéndez, M., Martinez, M. and Comín, F.A. (2001) A Comparative Study of the Effect of pH and Inorganic Carbon Resources on the Photosynthesis of Three Floating Macroalgae Species of a Mediterranean Coastal Lagoon. Journal of Experimental Marine Biology and Ecology, 256, 123-136. http://dx.doi.org/10.1016/S0022-0981(00)00313-0

61. Björnsäter, B.R. and Wheeler, P.A. (1990) Effect of Nitrogen and Phosphorus Supply on Growth and Tissue Composition of Ulva fenestrata and Enteromorpha intestinalis (Ulvales, Chlorophyta). Journal of Phycology, 26, 603-611. http://dx.doi.org/10.1111/j.0022-3646.1990.00603.x

62. Gressler, V., Yokoya, N.S., Fujii, M.T., Colepicolo, P., Mancini Filho, J., Torres, R.P. and Pinto, E. (2010) Lipid, Fatty Acid, Protein, Amino Acid and Ash Contents in Four Brazilian Red Algae Species. Food Chemistry, 120, 585-590. http://dx.doi.org/10.1016/j.foodchem.2009.10.028

63. Peters, K.J., Amsler, C.D., Amsler, M.O., McClintock, J.B., Dunbar, R.B. and Baker, J. (2005) A Comparative Analysis of the Nutritional and Elemental Composition of Macroalgae from the Western Antarctic Peninsula. Phycologia, 44, 453-463. http://dx.doi.org/10.2216/0031-8884(2005)44[453:ACAOTN]2.0.CO;2

64. Dawcznski, C., Schubert, R. and Jahreis, G. (2007) Amino Acids, Fatty Acids, and Dietary Fibre in Edible Seaweed Products. Food Chemistry, 103, 891-899. http://dx.doi.org/10.1016/j.foodchem.2006.09.041

65. Polat, S. and Ozogul, Y. (2008) Biochemical Composition of Some Red and Brown Macro Algae from the Northeastern Mediterranean

Sea. International Journal of Food Sciences and Nutrition, 59, 566-572. http://dx.doi.org/10.1080/09637480701446524

66.   Prabhasankar, P., Ganesan, P., Bhaskar, N., Hirose, A., Stephen, N., Gowda, L.R., Hosokawa, M. and Miyashita, K. (2009) Edible Japanese Seaweed, Wakame (Undaria pinnatifida) as an Ingredient in Pasta: Chemical, Functional and Structural Evaluation. Food Chemistry, 115, 501-508. http://dx.doi. org/10.1016/j.foodchem.2008.12.047

67.   Matanjun, P., Mohamed, S., Mustapha, N.M. and Muhammad, K. (2009) Nutrient Content of Tropical Edible Seaweeds, Eucheuma cottonii, Caulerpa lentillifera and Sargassum polycystum. Journal of Applied Phycology, 21, 75-80. http://dx.doi.org/10.1007/ s10811-008-9326-4

68.   Fleurence, J., Morançais, M., Dumay, J., Decottignies, P., Turpin, V., Munier, M., Garcia-Bueno, N. and Jaouen, P. (2012) What Are the Prospects for Using Seaweed in Human Nutrition and for Marine Animals Raised through Aquaculture? Trends in Food Science & Technology, 27, 57-61.http://dx.doi.org/10.1016/j. tifs.2012.03.004

69.   Munier, M., Dumay, J., Morançais, M., Jaouen, P. and Fleurence, J. (2013) Variation in the Biochemical Composition of the Edible Seaweed Grateloupia turuturu Yamada Harvested from Two Sampling Sites on the Brittany Coast (France): The Influence of Storage Method on the Extraction of the Seaweed Pigment R-Phycoerythrin. Journal of Chemistry, 2013, Article ID: 568548, 8 pages. http://dx.doi.org/10.1155/2013/568548

70.   Marinho-Soriano, E., Fonseca, P.C., Carneiro, M.A.A. and Moreira, W.S.C. (2006) Seasonal Variation in the Chemical Composition of Two Tropical Seaweeds. Bioresource Technology, 97, 2402-2406. http://dx.doi.org/10.1016/j.biortech.2005.10.014

71.   Yu, J. and Yang, Y.F. (2008) Physiological and Biochemical Response of Seaweed Gracilaria lemaneiformis to Concentration Changes of N and P. Journal of Experimental Marine Biology and Ecology, 367, 142-148.http://dx.doi.org/10.1016/j.jembe.2008.09.009

72.   Maurin, C. and Le Gal, Y. (1997) Glutamine Synthetase in the Marine Coccolithophorid Emiliania huxleyi (Prymnesiophyceae): Regulation of Activity in Relation to Light and Nitrogen Availability.

Plant Science, 122, 61-69. http://dx.doi.org/10.1016/S0168-9452(96)04539-6

73. Perfeto, P.N.M. (1998) Relation between Chemical Composition of Grateloupia doryphora (Montagne) Howe, Gymnogongrus griffithsiae (Turner) Martius, and Abiotic Parameters. Acta Botanica Brasilica, 12, 77-88.

74. Renaud, S.M. and Luong-Van, J.T. (2006) Seasonal Variation in the Chemical Composition of Tropical Australian Marine Macroalgae. Journal of Applied Phycology, 18, 381-387. http://dx.doi.org/10.1007/s10811-006-9034-x

75. Khairy, H.M. and El-Shafay, S.M. (2013) Seasonal Variations in the Biochemical Composition of Some Common Seaweed Species from the Coast of Abu Qir Bay, Alexandria, Egypt. Oceanologia, 55, 435-452. http://dx.doi.org/10.5697/oc.55-2.435

76. Fleurence, J. (1999) Seaweed Proteins: Biochemical, Nutritional Aspects and Potential Uses. Trends in Food Science & Technology, 10, 25-28.http://dx.doi.org/10.1016/S0924-2244(99)00015-1

77. Rowan, K.S. (1989) Photosynthetic Pigments of Algae. Cambridge University Press, New York.

78. Payri, C.E., Maritorena, S., Bizeau, C.E. and Rodière, M. (2001) Photoacclimation in the Tropical Coralline Alga Hydrolithon onkodes (Rhodophyta, Corallinaceae) from a French Polynesian Reef. Journal of Phycology, 37, 223-234. http://dx.doi.org/10.1046/j.1529-8817.2001.037002223.x

Applicative Study (Part I): The Excellent Conditions to Remove in Batch Direct Textile Dyes (Direct Red, Direct Blue and Direct Yellow) from Aqueous Solutions by Adsorption Processes on Low-Cost Chitosan Films under Different Conditions

Vito Rizzi[1], Alessandra Longo[1], Paola Fini[2], Paola Semeraro[1], Pinalysa Cosma[1, 2], Esther Franco[3], Rocío García[3], Marcela Ferrándiz[3], Estrella Núñez[4], José Antonio Gabaldón[4], Isabel Fortea[4], Enrique Pérez[5], and Miguel Ferrándiz[5]

[1]Università degli Studi "Aldo Moro" di Bari, Dip. Chimica, Bari, Italy

[2]Consiglio Nazionale delle Ricerche CNR-IPCF, UOS Bari, Bari, Italy

[3]Biotechnology Department, Textile Industry Research Association (AITEX), Alcoy, Spain

[4]Departamento Ciencia y Tecnología de Alimentos, Universidad Católica San Antonio de Murcia, Guadalupe, Murcia, Spain

[5]Colorprint Fashion, SL, Avda. Fco. Vitoria Laporta, Muro de Alcoy (Alicante), Spain

ABSTRACT

In recent years the development of chitosan (CH) based materials as useful adsorbent polymeric matrices is an expanding field in the area of adsorption science. Even though CH has been successfully used for dye removal from aqueous solutions due to its low cost, no considerations have been made about, for example, the effect of changing the pH of chitosan hydrogelor about the dehydrating effect of Ethanol (EtOH) treatment of chitosan film on the dyes removal from water. Consequently in our laboratory we carried out a study focusing the attention, mainly, on the potential use of CH films under different conditions, such as reducing the intrinsic pH, increasing the hydrophobic character by means of ethanol treatment and neutralization of CH films to improve their absorption power. Textile anionic dyes named Direct Red 83:1, Direct Yellow 86 and Direct Blue 78 have been studied with the aim of reducing the contact time of CH film in waste water improving the bleaching efficiency. Neutralized acid CH film and longtime dehydrated one result to be the better films in dye removal from water. Also the reduction of the CH solution acidity during the film preparation determines the decreasing of the contact time improving the results. The effect of initial dye concentration has been examined and the amount of dye adsorption in function of time t, q_t (mg/cm^2), for each analyzed film has been evaluated comparing the long term effect with the decoloration rate. A linear form of pseudo-first-order Lagergren model has been used and described. The best condition for removing all examined dyes from various dye solutions appears to be the dehydration of a novel projected CH film obtained by means of the film immersion in EtOH for 4 days. Also CH films prepared by well-known literature procedure and neutralized with NaOH treatment appear having an excellent behavior, however the film treatment requires a large quantity of water and time.

INTRODUCTION

Dyes and Wastewater Treatments

Water represents the most very important human resource with economic, social, political and environmental importance throughout the world [1]. Recently, different pollutants are inflowing aquatic systems since the rapid industrialization and urbanization [2]. Considerable volumes of water were consumed by textile, paper, plastics and dyestuffs industries due to the use of very different chemical reagents during manufacturing and coloring of their products. Subsequently, they produce a considerable amount of polluted wastewater and among them color is the first contaminant to be recognized in wastewater and the presence of very small amounts of dyes in water is highly visible and undesirable [3]. Thus, every year, tons of dyes are discharged from textile, pharmaceutical, packed food, pulp and paper, paint, plastics, petroleum, electroplating, and cosmetics industries into the environment worldwide [4]. This together with a strict legislation on the discharge of these toxic products in the environment makes necessary to develop various efficient technologies for the removal of pollutants from wastewater [5] with the result that the treatment of wastewater becomes a matter of dominant importance. Efforts to develop processes for removing assorted pollutants have mainly attempted by researchers in the analytical, environmental and material sciences and more recently different technologies and processes are used. Extensively used technology for the removal of both inorganic and organic material could be classified in biological treatments, membrane processes, advanced oxidation processes, chemical and electrochemical techniques [6], and adsorption procedures [7]. In spite of the availability of different proposed treatments, the physical adsorption by means of sorbent materials represents one of the most fashionable methods since proper design of the adsorption process will produce high-quality treated effluents. Physical adsorption is a well-known equilibrium separation process and since it is widely recognized that adsorption using low-cost adsorbents is an useful and economic method for wastewater decontamination, a number of materials have been extensively investigated [5].

Chitosan: a Low Cost Dyes-Bioadsorbent

Recently the attention has been focused on a variety of biosorbent materials [8] [9]. Due to their low cost the use of adsorbents composed of natural polymers has attracted significant interest, and polysaccharides such as Chitosan (CH) and its derivatives have received great attention. CH is an effective adsorbent compared with activated carbons and other common adsorbents used in treatment of organic or inorganic contaminated water [10]. It is a derivative of chitin, one of the most abundant natural polymers in the biosphere, that represent the main component of the exoskeletons of marine crustaceans (e.g., shrimps, crabs, krill), which are available in large amounts as a byproduct of food processing [11]. Depending on the source and preparation procedure, the number of glucosamine residues (denoted as the degree of deacetylation) on the linear polysaccharidic chain of CH is above 60% [12]. CH is considered a super high-capacity adsorbent for contaminant removal from water due to its ability of binding contaminants through hydroxyl and amino groups on the surface [13] . In particular due to the protonation of the primary amino groups of CH by acidic media, it can strongly adsorb anionic dyes such as acid, reactive and direct dyes, by electrostatic attraction [14] [15] and also metal anions by ion exchange. Since pH variation leads to the difference in the degree of ionization of the adsorptive molecule and the surface properties of adsorbent, the pH of the medium is one of the most important factors affecting the capacity of adsorbent in wastewater treatment and consequently the efficiency of adsorption is strictly related on the media pH [16] . It is worth noting that the amount of dye removal by means of adsorption processes is greatly related to the chemical nature [17] (For more details see Ref. [18]) and to the initial concentration of the dyes. Concerning this aspect several ways for classification of commercial dyes are known in literature and generally they can be classified in terms of chemical structure, color and application method [19] . However, due to the complexities of the color nomenclature from the chemical structure system, the classification based on application is often considered more useful [20] . A classification based on their particle charge upon dissolution in aqueous application medium such as cationic (all basic dyes), anionic (direct, acid, and reactive dyes), and non-ionic (dispersed dyes) has been used yet [18] . Since the increasing number of review articles published on the use of CH and its derivatives as adsorbents for the removal of contaminants from wastewaters [6] in the present paper we summarize the ability of modified CH films in binding different textile dyes, in particular azoic ones, considered extremely toxic due to the presence of toxic amines as by-products in the effluent under varying conditions [18] [21] - [23] It has been shown the modified procedure to prepare CH films and the sorption mechanism related to the dye decoloration rate. The attention has been focused on three direct dyes, Direct Red 83:1 (DR), Direct Yellow 86 (DY) and Direct Blue 78 (DB), analyzing the pH effect both of the medium and of the film on dye uptake from aqueous solutions. Also the influence of neutralization and increasing hydrophobicity of CH films has been examined. The obtained results indicate that this novel modality of producing CH films allows the treatment of a large amount of wastewater which can be purified with an efficiency around of 100% in a very short period of time using less energy and low cost procedure.

MATERIAL AND METHODS

All the chemicals used were of analytical grade and samples were prepared using double distilled water. Commercial grade CH powder (from crab shells, highly viscous, with deacetylation degree $\geq$ 75%), Acetic acid (99.9%), Ethanol (EtOH) (99.9%) and glycerol (+99.5%) were purchased from Sigma-Aldrich. Direct Red 83:1, Direct Yellow 86 and Direct Blue 78 were received by Colorprint Fashion, S.L within the LIFE+ European Project named "DYES4EVER" (Demonstration of Cyclodextrin Techniques in Treatment of Waste Water in Textile Industry to Recover and Reuse Textile Dyes) and used without further purification. Direct dyes characteristics are: Direct Red (Color Index Number: 83:1), chemical formula: $C_{33}H_{20}N_6Na_4O_{17}S_4$, MW: 992.77 g mol^{-1}; Direct Yellow (Color Index Number: 86), chemical formula: $C_{39}H_{30}N_{10}Na_4O_{13}S_4$, MW: 1066.94 g mol^{-1} and Direct Blue (Color Index Number: 78), chemical formula: $C_{42}H_{25}N_7Na_4O_{13}S_4$, MW: 1055.1 g·mol^{-1}, (see Scheme 1).

Dye stock solutions with a concentration of 10^{-4} M were prepared and dilutions were carried out with double distilled water to obtain different concentrations, 5×10^{-5} M and 5×10^{-6} M.

The pH of the various aqueous solutions has been adjusted using concentrated HCl and NaOH solutions.

UV-Vis absorption spectra were recorded in the 200 - 800 nm range, at a 0.5 nm/s scan rate, using a Shimadzu UV-Vis spectrophotometer mod. 1601.

Preparation of Chitosan Films

The standard procedure used in literature to prepare CH films [24] [25] was adopted. In order to optimize the adsorption processes, pH medium effect and dehydration of CH films by means of EtOH were studied, too. CH powder was dissolved in 0.8% (v/v) aqueous acetic acid solution in order to obtain a 1% (w/v) chitosan concentration by constant continuous stirring for 24 hrs. 200 µL of glycerol were added every 100 mL of CH acetic solution. Then the solution was filtered through a coarse sintered glass filter and degassed for 1 hr. After degassing, the CH solution was poured into a plastic Petri plate. This plate was maintained in an oven at 60°C for 24 hrs. A thin CH membrane was obtained.

(a)

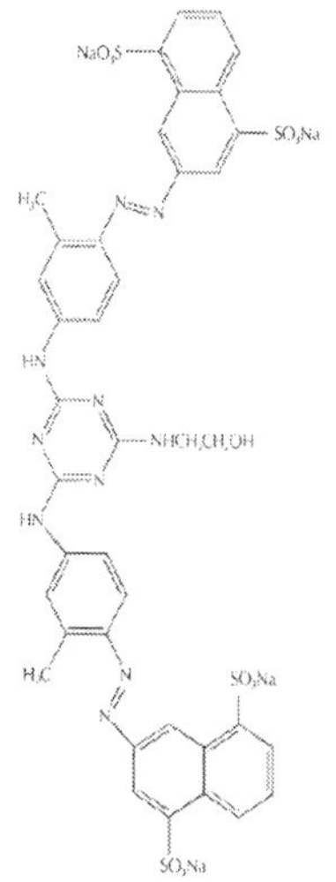

(b)

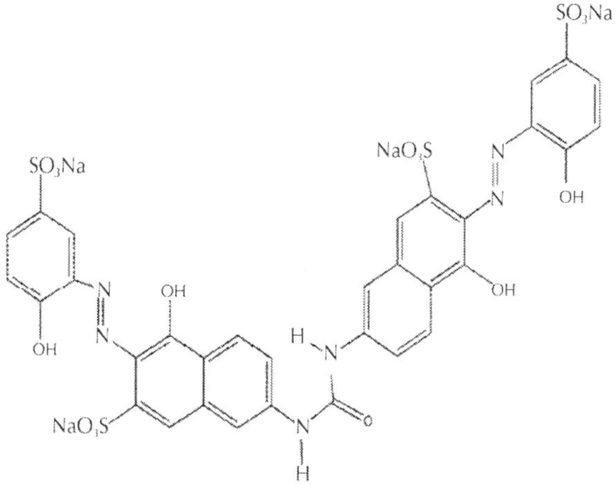

(c)

Scheme 1: Molecular structure of used dyes: Direct Blue 78 (a); Direct Yellow 86 (b) and Direct Red 83:1 (c).

In order to reduce the intrinsic acidity of CH hydrogel medium and to low the total cost to prepare CH films, a novel procedure was adopted modifying the standard procedure known in literature. In particular an excess of CH powder was used. 2 g were solubilized in 100 mL of known acetic acid solution (0.1% v/v) containing the same amount of glycerol (200 µL every 100 mL acetic solution). In this way the CH solution was almost completely neutralized and pH 6.2 was established. As indicated in the previous procedure, the solution was filtered, however in this case a great amount of CH powder remains insoluble, consequently could be recovered and potentially used to prepare new films reducing the total cost. Small amount of CH was required to realize the membrane. The obtained free standing films were sliced in order to obtain 2.5 cm × 2.5 cm films.

About the CH films obtained according to the standard procedure illustrated in Krajewska et al. [24] [25] , the effect of film neutralization with a NaOH solution (0.2 M) and the effect of the dye solutions pH on dye adsorption processes, were studied too. Concentrated NaOH and HCl were employed to adjust the pH of dye solutions, establishing pH 12 and 2, respectively.

Neutralization of acid CH films was obtained immerging the film in the 0.2 M NaOH solution (40 mL) for 30 min. Successively the film was washed by immersion in double distilled water for six days changing water every day to ensure the complete neutralization. The effect of EtOH treatment was estimated immerging prepared CH films in EtOH (40 mL) for 24 h and 96 h.

Batch Experiments and Adsorption Isotherms

Adsorption kinetics were carried out by in batch technique. Batch experiments were carried out for determining the adsorption isotherms of dyes onto the various adsorbent CH films in a glass beaker. The dye aqueous solutions were magnetically stirred at a constant rate (140 rpm), allowing sufficient time for reaching adsorption equilibrium. It was assumed that the applied stirring speed allows all the film surface area to come in contact with dye molecules over the course of the experiments. In each experiment a fixed volume (40 mL) of dye aqueous solution at constant dye concentration (5×10^{-5} M and 5×10^{-6} M) was used. The study was performed at room temperature to be representative of environmentally relevant condition. All experiments were carried out in duplicate and the average value was used for further calculation.

For the adsorption kinetic study, sample aliquots were withdrawn from the flask at predetermined time intervals and the UV-Vis absorption spectrum was recorded. Measurements were carried out for several days until the equilibrium was reached, however to estimate the adsorption rate constants only data collect in the first 250 min were employed. According to the procedure used in Run Fang et al. [26] , it is possible to evaluate the efficiency on equilibrium of adsorption process using the following decoloration rate equation (Equation (1)):

$$E = \frac{A_0 - A}{A_0} \times 100$$

(1)

where A_0 and A represent the intensity of the dye absorption band at a fixed wavelength, at time zero and at equilibrium time, respectively. The amount of adsorbed dye at a certain time t, q_t(mg/cm^2) was calculated by Equation (2):

$$q(t) = \frac{C_0 - C_t}{A_{CH}} \times V$$

(2)

where V represents the total volume of solution (herein 40 mL), A_{CH} is the area of the analyzed CH films (obtained redoubling the geometrical area of the films), C_0 and C_t represent the initial concentration and the concentration at time t for each dye. The A_{CH} was calculated by the following formula (Equation (3)):

$$A_{CH} = 2(d \times d)$$

(3)

in which d represent the size of the film side. In this case the film side measures about 2.5 cm, then the total superficial area is about 12.5 cm^2. To calculate the mass of adsorbed dye in milligrams, the epsilon values used are reported in Table 1, expressed in M^{-1}×cm^{-1} units.

The time evolution of the dye adsorption was interpreted by means of a pseudo-first-order model described by Lagergren in the form:

Table 1: Comparison of the pseudo-first-order, pseudo-second-order adsorption rate constants and calculated and experimental q_e values obtained at different Direct Yellow 86 concentrations

Film	$q_{e,th}$ (mg/cm^2)	$q_{e,exp}$ (mg/cm^2)	Pseudo-first order			Pseudo-second order		
			$q_{e,calc}$ (mg/cm^2)	K (min^{-1})	R^2	$q_{e,calc}$ (mg/cm^2)	K (min^{-1})	R^2
DY/CH new								
Connected dye								
CH	0.17	0.16	0.14	(35.00 ± 2.00) × 10^{-4}	0.96	0.12	(1.32 ± 1.00)	0.99
CH/EtOH 1d	0.17	0.17	0.16	(9.00 ± 1.00) × 10^{-4}	0.96	0.07	(1.22 ± 1.00)	0.99
CH/EtOH 4d	0.17	0.16	0.16	(3.00 ± 1.5) × 10^{-4}	0.69	0.022	/	/
Diluted dye								
CH	0.019	0.014	0.015	(20.00 ± 1.00) × 10^{-4}	0.97	3.7	/	/
CH/EtOH 1d	0.019	0.016	0.014	(43.00 ± 4.00) × 10^{-4}	0.90	0.024	(81.32 ± 56.35) × 10^{-2}	0.89
CH/EtOH 4d	0.019	0.016	0.012	109.00 ± 7.00) × 10^{-4}	0.94	0.029	(1.01 ± 0.45)	0.89
DY/CH std								
Connected dye								
CH pH12	0.17	0.17	0.12	22.00 ± 1.00) × 10^{-4}	0.98	0.09	(1.14 ± 0.08)	0.98
CH/NaOH	0.19	0.24	0.24	40.00 ± 1.00) × 10^{-4}	0.99	0.15	(7.00 ± 2.00) × 10^{-3}	0.98
CH/EtOH 1d	0.19	0.17	0.26	42.00 ± 1.00) × 10^{-4}	0.98	0.13	53.09 ± 14.88) × 10^{-3}	0.98
CH/EtOH 4d	0.20	0.21	0.20	20.00 ± 1.00) × 10^{-4}	0.98	0.13	62.28 ± 25.00) × 10^{-3}	0.98
Diluted dye								
CH pH12	0.019	0.019	0.018	24.00 ± 1.50) × 10^{-4}	0.97	0.0013	62.10 ± 18.40) × 10^{-3}	0.99

CH/NaOH	0.020	0.026	0.025	81.00 ± 1.00) × 10⁻⁴	0.99	0.0036	19.61 ± 4.66) × 10⁻³	0.99
CH/EtOH 4d	0.020	0.022	0.0021	50.00 ± 1.50) × 10⁻⁴	0.99	0.0023	35.65± 15.04) × 10⁻³	0.94

$$\text{Log}\left(q_e - q_t\right) = \text{Log}q_e - \frac{Kt}{2.303}$$

where q_e and q_t are the adsorption capacity (mg/cm^2) at equilibrium and at time t (min), respectively; k (min^{-1}) is the adsorption rate constant of pseudo-first-order and was calculated from the linear plot of log(q_e – q_t) versus t [18] .

RESULTS AND DISCUSSION

The UV-Vis absorption spectra of all studied dyes recorded in aqueous solution at different pHs are reported in Figure 1. The absorption spectra are dominated by the classical absorption bands relative to the characteristics chromophore for each dye, result of the interaction between azo functionality (–N=N–) and attached aromatic moieties. In general there are weaker bands in the UV region attributable to the electronic transitions related to the aromatic rings, while in the visible range there are more intense and wider bands due to ϖ ϖ^* transitions of do nor groups, e.g. an aromatic nucleus containing an auxochromic group such as alkyl side chains, secondary amine, or OH group [27] - [29] . The characteristic absorption bands in aqueous solution at pH7 are located at 605 nm for DB, at 365 nm for DY and at 527 nm for DR. The peak position changes varying the pH of aqueous solution only for DY and DR, while results to be unaffected by changing solution pH in the case of DB. In particular for acid and basic pH the DY absorption band shifts to 390 nm, whereas the DR absorption band at acid pH shifts to 553 nm and for basic pH is located at 535 nm. The decoloration rate has been evaluated monitoring the changes in time of the dye characteristic absorption bands in the visible range spectrum varying the CH film type and the dye solution properties. In general, for all examined cases, the efficiency of dye adsorption depends on the pH of the aqueous solution in which the CH film is immersed influencing not only the surface charge of the adsorbent (due to the presence of functional group such as NH_3^+ group on CH film surface), the degree of ionization of the material present in the solution and the dissociation of functional groups on the active sites of the adsorbent, but also the solution dye chemistry [14] [18] .

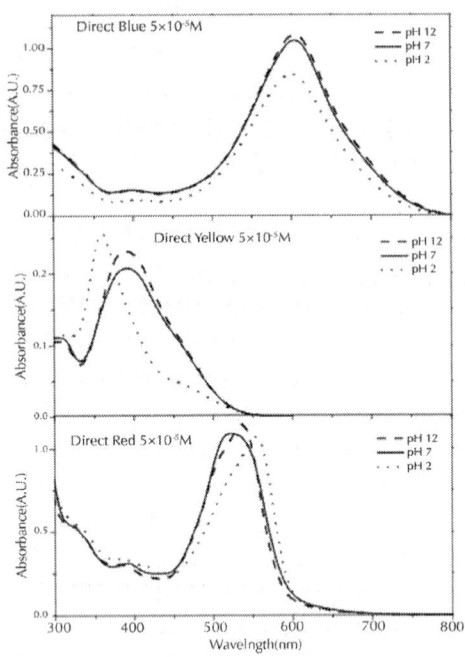

Figure 1: UV-VIS absorption spectra of aqueous solutions of dyes at different pH.

In Figures 2 (panel a-f) the comparison between standard CH film (CH std), prepared following the classical method, and CH film (CH new), prepared with the new method, about the efficiency in bleaching dye solution for each dye is reported. Traditional CH films have the property to swell in a large way when immersed in aqueous acidic solution compared with the same CH films both when it is immersed in basic (pH 12) aqueous solution or in EtOH, and when it is neutralized by NaOH treatment [30] . In particular CH films treated with an acid dye solution swell in large way because of the electrostatic repulsion among

CH protonated NH_2 groups resulting in a reduced dyedecoloration rate which cannot be evaluated. Further it is known that an excess of acidity, considering a pH < 2, induces a strong competition of theanions, brought about by the dissociation of the acid used for pH control, and the sorption capacity significantly decreased [31] . It is also possible to observe the precipitation of almost all examined dyes due to their lower solubility at acidic pHs [20] . Consequently CH std and CH std film treated with an acid solution (CH pH 2) appear useless for all considered dyes. On the contrary better results were obtained for traditional CH films treated with basic solutions (CH pH 12), with EtOH (CH/ETOH) and neutralized with NaOH (CH/ NaOH) solution. This behavior could be attributed to the complete or partial neutralization of NH_3^+ groups present on chitosan chains. Also the initial dye concentration is considered an important factor. The increase of the initial dye concentration determines an increase in the adsorption capacity of the adsorbent and this may be due to the high driving force for mass transfer at a high initial dye concentration [17] . A plateau region is established after some time and this plateau becomes much more evident for concentrated dye solution. In Figure 2(a) and Figure 2(b) are reported the decoloration rates referred to DY for CH std and CH new films respectively, after equilibrium has been reached. Surprisingly in all examined case spectroscopic data show the almost complete uptake of dye onto the different CH film. Better results are obtained for neutralized and long time dehydrated films. About the effect of initial dye concentration, in general the observed behavior appears to reflect the classical trend in which the amount of the dye adsorbed onto chitosan film increased with the increase of the initial concentration of dye solution under the same adsorbent amount. The only system for which it was not possible to verify this trend was the CH std after one day of dehydration (CH/ETOH 1d), since the film did not reach any equilibrium condition. About the time required for reaching the highest DY decoloration efficiency there is a wide range of possibility from 7 hrs (relative to the neutralized CH film at the lower initial dye concentration) until to 17 days in the case of new type of CH film without any treatment. In general the new CH films (Figure 2(b)) seem to need a lower contact time with the exception of the CH film used without modification which reaches the equilibrium in 17 days. The effect of dehydration treatment of CH films, realized by immersion of the film in EtOH for different times, results to be very interesting.

In this case in 1 or 2 day occurs the complete removal of dye from both concentrated and diluted dye solutions, respect to the 4 or 5 days necessary for CH std in the same conditions. However, in spite of the long equilibrium time (17 days and 3 days for concentrated and dilute dye solution, respectively) needed for CH new in dye solution, those films show exceptional results due to the absence of collateral effects such as, for example, the acidification of the stained aqueous solution which always occurs using CH std films. Of course, in term of equilibrium time and 100% decoloration rate, treated CH new, in particular the film dehydrated 4 days (CH/ETOH 4d), appears to be the best in its ability to remove DY dye from dilute and concentrated solutions. Furthermore, although the better adsorption characteristic are shown by CH/NaOH films, the use of CH new films, dehydrated and not dehydrated with EtOH, result more interesting because no further water washing cycles or neutralization are required and the long term absorption time ensure anyway an efficient decoloration of dyed solution.

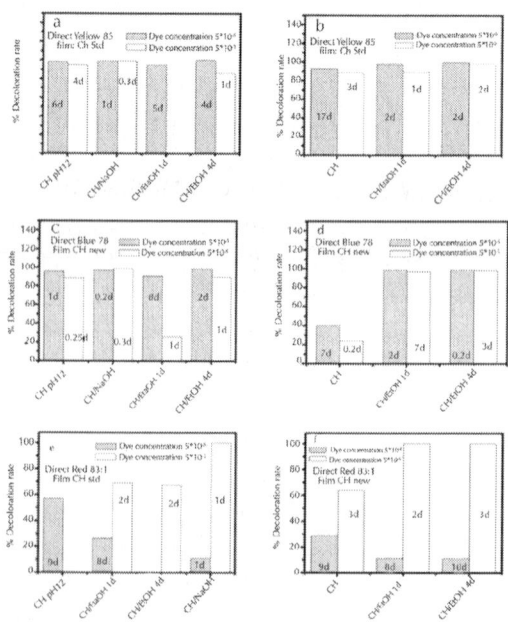

Figure 2: Comparison of decoloration rate percentage for different dyes in presence of CH films. (a) DY/CH std; (b) DY/CH new; (c) DB/CH

std; (d) DB/CH new; (e) DR/CH std; (f) DR/CH new. Dense pattern is referred to highest dye concentration 5×10^{-5} M, white columns to lowest dye concentration 5×10^{-6} M.

In the Figure 2(c) and Figure 2(d) results related to DB dye are reported. Generally speaking, the CH film behavior appear very similar to the that shown in presence of DY, even though shorter equilibrium time is experimented in obtaining the same decoloration efficiency. In the case of CH std film the maximum of decoloration rate is reached for the neutralized CH std film (CH/NaOH), for long term dehydrate film (CH/ETOH 4d) and for film in alkaline dye solution (CH pH 12). As occurs in the case of DY, also for DB the decolorationrate collapses in acidic medium and, for long contact time, a dye precipitation has been also observed both for untreated CH std film and CH std film immersed in acid dye solution. The results obtained for CH std film previously neutralized in NaOH solution appear interesting. Only few hours are needed to completely remove the dye from solution both for concentrated and dilute solution. In the Figure 2(d) the effect of CH new film has been shown and not surprisingly the long time dehydrate film seems the best in term of decoloration rate. Unlike the system DY/CH new film, in this case the untreated CH new film appears the worst and a strong reduction of the decoloration rate has been observed. About the initial dye concentration effect it can observe that, reducing the dye concentration, the equilibrium time for an efficient removal of DB increments for all analyzed films. This could be attributed to a decrease in the driving force of the concentration gradient with the reduction of the initial dye concentration [14] . This effect is particularly evident for CH/ETOH 1d and 4d in which, decreasing the dye concentration, the time needed to efficiently remove the dye from aqueous solution increases very much passing from 2 days to 7 days for the CH film partially dehydrated and from 5 h to 3 days for the CH film completely dehydrate. However, as observed in the previous case, although the CH/NaOH film seems to be the best in terms of contact time and efficiency in decoloration rate for both DB aqueous solutions, the use of CH/ETOH 4d film results better avoiding the wash water cycle, which is time and water consuming, and the time needed to neutralize the film. In the panels e and f of the Figure 2 the results related to DR dye in presence of various CH films are reported. It is evident that the DR adsorption onto CH films is not so efficient for both examined film typologies (traditional and new method of film preparation) at high initial dye concentration. In particular only in the case of the standard CH film immersed in a pH 12 dye aqueous solution, the system appears to be able to adsorb the dye with a relatively good efficiency at an initial DR concentration of 5×10^{-5} M. The standard CH pH 12 film can remove about the 60% of the dye, after 9 days. Conversely for diluted dye solutions good results have been obtained with an equilibrium time of 2 - 3 days for both CH film typologies. Among the CH films prepared by means of the standard procedure, the CH film treated with NaOH seems to be the more functional system. While in the case of CH film prepared with the new method, the change in the film procedure preparation ensure the same efficiency after 2 or 3 days if CH new film was previously treated with EtOH. As in the case of DB and DY, new CH/ETOH films show an excellent performance in removing DR dye with a decoloration rate of 100% provide that the dye concentration is maintained under 10^{-5} M.

As it is evident equilibrium time for reaching the highest decoloration rate is very different for the 3 studied dyes and it appears related to the dimensions and concentration of dyes, besides to the interactions between adsorbent film and dye molecules involving both electrostatic and Van der Waals interactions [26]. Although in general, the adsorption capacity increases with time, at certain point the system reaches a state of dynamic equilibrium between the amount of dye being adsorbed onto the material and the amount of dye desorbed from the adsorbent, then it is important to optimize this contact time, considering the efficiency of desorption and regeneration of the adsorbent. In the case of DY and DB, increasing initial dye concentration the result seems to get better although long contact times are required. Instead, decreasing the initial concentration it is observed a decrease of equilibrium time, together with a efficiency reduction around the 90% in almost all the DY systems, while better results have been achieved for DB. Only in the case of standard CH/NaOH and CHnew/ETOH 4d the effect in term of decoloration efficiency appear to be unrelated with the dye concentration, for both dyes. This different behavior could be interpreted considering that, in general, during the adsorption process, the dye molecules progressively occupy the active sites present on the adsorbent surface, covering the film surface after some time and forming a monolayer dye coverage on the outer surface of the film. Obviously a large number of vacant surface sites are available for adsorption during the initial stage, and after a lapse of time, it is difficult to occupy there maining vacant surface sites due to repulsive forces between dye molecules adsorbed on the solid and those in the solution phase. Then after the dye adsorption from the solution to the adsorbent surface, the last stage of the adsorption process could be related to the diffusion of the dye-adsorbed molecules within the material. This last step can be strictly correlated to the sizes of the dye organic chains, to the amount and positioning of the sulfonate groups of the dyes and to the adsorption temperature [19] [31] -[33] . About this aspect, the adsorption from lower dye concentration appears to be more efficient for almost all analyzed dyes indicating that probably the rate determining step in reaching the equilibrium time is a combination between the rate in the active site occupation, which is directly correlated to the initial dye concentration [19] , and the adsorbed dye molecules diffusion into the film, which is correlated to the dye molecular size and number of charged groups able to electrostatically interact with charged group on chitosan film [34] .

Concerning kinetic treatments, as already said, the effect of initial dye concentration depends on the direct relation between the concentration of the dye and the available sites on the adsorbent surface for both typologies of film (CH std and new) and for all analyzed dye. In general the behavior of the Direct Yellow and Direct Blue is quite similar, while Direct Red differentiates. In theFigure 3 and Figure 4 (panel a and c) the adsorption profiles of the DY (Figure 3(a) and Figure 3(c)) and DB (Figure 4(a) and Figure 4(c)) in relation to the contact time are reported considering the two typologies of CH films. It is clear that the dye molecules progressively occupy the active sites of various CH films and, increasing the initial concentration of dye solution the adsorption capacity of CH films increases with time due to the driving force of the concentration gradient at least in the case of the higher dye concentration [19] . Furthermore it is also evident that equilibrium is reached in higher times for high different results have been obtained in the case of DR. In Figure 5 (panels a and c) the adsorption profiles values of initial dye concentration.

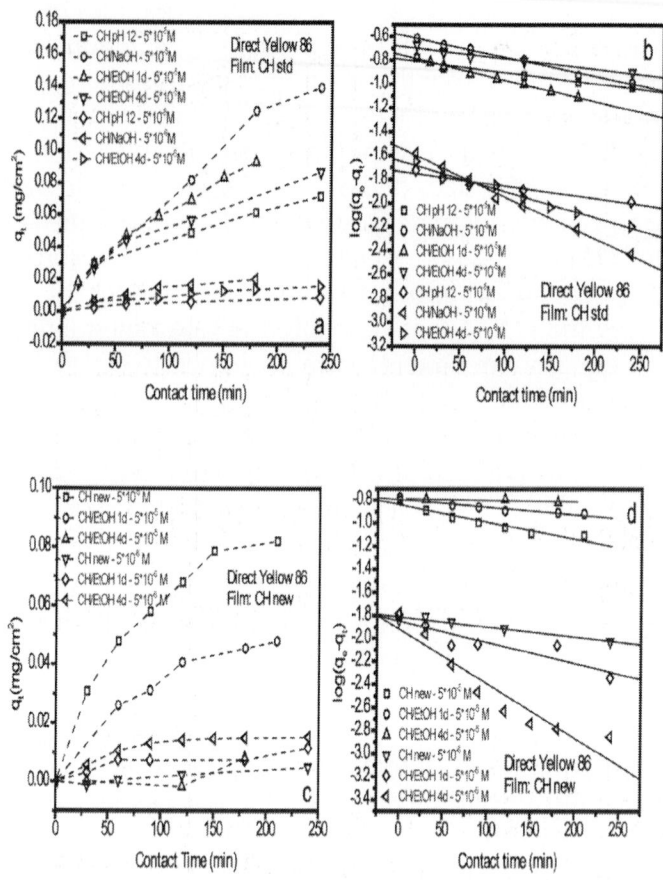

Figure 3: Amount of adsorption at time t, q_t (mg/cm²) versus time for DY in presence of CH std film (a) and CH new fil (c). Linearized first order plot for the absorption of DY in presence of CH std film (b) and CH new film (d).

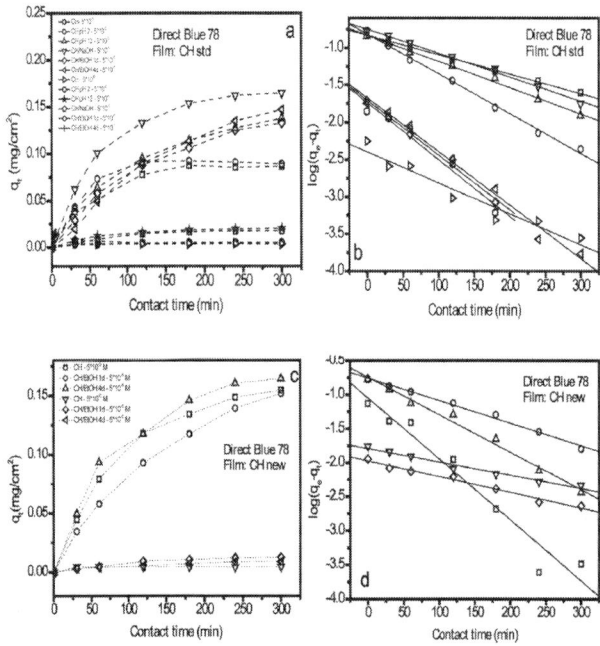

Figure 4: Amount of adsorption at time t, q_t (mg/cm²) versus time for DB in presence of CH std film (a) and CH new fil (c). Linearized first order plot for the absorption of DB in presence of CH std film (b) and CH new film (d).

In the process of dye adsorption, initially the dye molecules have to be transferred at the adsorbent surface and then finally, they have to diffuse into the porous structure of the adsorbent [35] . Therefore, dye at higher initial concentrations will take relatively longer contact time to attain equilibrium due to higher amount of dye molecules respect to the number of available surface adsorption sites that in time decrease. Further another important parameter, which could affect the contact time for reaching the equilibrium, is the dye diffusion rate into the adsorbent pores. Thus the removal of dyes depends on the dye initial concentration and on the chemical characteristics of the dye [19] [31] [33] [34] . The amount of adsorbed dye at the equilibrium time reflects the maximum adsorption capacity of the adsorbent under chosen operating conditions. In particular, in the case of CH std film (Figure 3(a) and Figure 4(a)), the q_t results to be very high for CH/NaOH film and decreases diluting 10 times the solution for both dyes. About the CH new film the dye behavior differs more. In fact, while it is possible to individuate a marked difference in the trend of the q_t at high and low initial dye concentration for DB (Figure 4(c)), in the case of DY the differences are less evident (Figure 3(c)). The untreated CH new film (CH) presents the highest dye uptake during the adsorption first phase for the DY (Figure 3(c)). For DB (Figure 4(c)), the CH/ETOH film shows the higher q_t. This could indicate that, after an initial relatively rapid dye adsorption under the dye concentration gradient (initial dye concentration effect), the saturation of the CH film vacant sites takes place slowing down the further dye adsorption in function of different chemical structure of the two dyes. The rapid initial phase is different for the two analyzed dyes and it may last from several minutes to few hours, while the following slow stage may continue for hours or days reaching a plateau. The behavior of system CH film/dye appears very interesting at low dye concentration where the plateau is absent. This can be interpreted considering that at low dye concentration the ratio of initial number of dye molecules to the active sites on the adsorbent film is low and subsequently the fractional adsorption becomes independent from initial dye concentration.

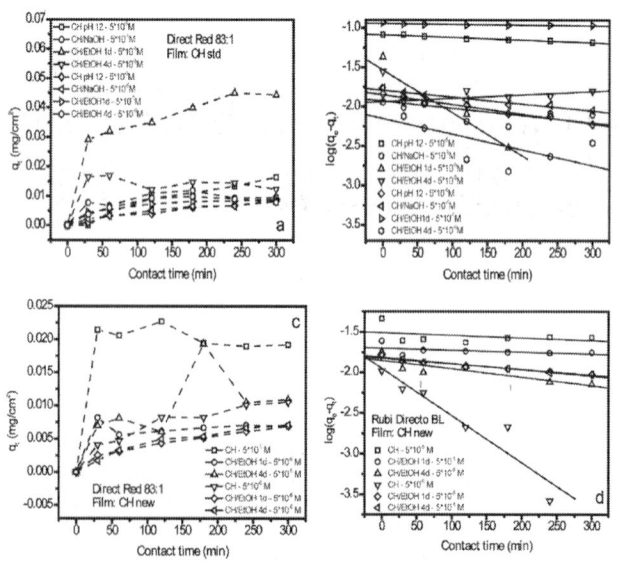

Figure 5: Amount of adsorption at time t, q_t (mg/cm^2) versus time for DR in presence of CH std film (a) and CH new fil (c). Linearized first order plot for the absorption of DR in presence of CH std film (b) and CH new film (d).

Different results have been obtained in the case of DR. In Figure 5 (panels a and c) the adsorption profiles of the red dye in relation to the contact time for various CH films are reported. It is evident that there are no marked variation of the q_t with the initial dye concentration for both examined film typologies, as already observed for the decoloration rate (panel e and f in Figure 2). For the CH std film type (Figure 5(a)), the CH film treated with ETOH for 1 day (CH/ETOH 1d) shows the highest dye uptake for a dye concentration of 5×10^{-5} M. While, for the same dye concentration, CH film without further treatment (CH in Figure 5(c)) results to have the highest adsorption capacity within the CH films prepared with the new method. No significant differences have been observed for the other film types in function of the initial dye concentration indicating that the effect of mass transfer could be considered irrelevant for the system DR/CH film. In order to investigate the mechanism of adsorption, a pseudo-first order and a pseudo-second order adsorption model has been used to test dynamical experimental data. The pseudo-first order and pseudo-second order kinetic models assume that the adsorption is a pseudo-chemical reaction. Many kinetic models have been developed with the aim to find intrinsic kinetic adsorption parameters [19] . Traditionally, data on kinetics of adsorption are described following the expressions originally given by Lagergren, which has adapted the general Langmuir rate equation to special cases [36] [37] . In Figures 3-5 (panels b, for CH std and d, for CH new) the trends of adsorption process calculated by means of the Lagergren model are reported for the different CH film typologies. An ideal adsorbent for waste water pollution remediation must not only present a large adsorbate capacity but also a fast adsorption rate. Therefore, the adsorption rate is another important parameter for selecting the adsorbent material and adsorption kinetics have to be considered since they explain how fast the chemical reaction occurs and also provides information on the factors affecting the reaction rate. About the kinetics of adsorption, differences in chitosan type, preparation, dyes and methodology examined makes any comparison of results difficult [19] . Table 1, Table 2 and Table 3 show the values of the pseudo-first order and pseudo-second order models together with the coefficient of correlation (R^2) and the average relative error for the three analyzed dyes in relation to the different types of CH films. Assuming a pseudo-first order kinetics, the calculated q_e ($q_{e,calc}$) values are agree with the experimental q_e values. This suggests that the adsorption of dye follows a first-order kinetics for all analyzed CH films. Good correlation coefficients have been also obtained for almost all examined dyes. In the case in which dye have been completely absorbed, the great efficiency shows by the used CH films is demonstrated by the comparison between the experimental $q_{e,exp}$ and the theoretical values of expected $q_{e,th}$ calculated applying the following equation:

$$q_{e,th} = \frac{C_0}{A_{CH}} \times V$$

Table 2: Comparison of the pseudo-first-order, pseudo-second-order adsorption rate constants and calculated and experimental q_e values obtained at different Direct Blue 78 concentrations

Film	$q_{e,th}$ (mg/cm²)	$q_{e,exp}$ (mg/cm²)	Pseudo-first order			Pseudo-second order		
			$q_{e,calc}$ (mg/cm²)	K (min⁻¹)	R²	$q_{e,calc}$ (mg/cm²)	K (min⁻¹)	R²
DB/CH new								
Connected dye								
CH	0.16	0.074	0.089	(2.00 ± 0.1) × 10⁻²	0.97	0.21	(4.66 ± 1.08) × 10⁻²	0.99
CH/EtOH 1d	0.17	0.17	0.17	(8.00 ± 0.1) × 10⁻³	0.99	0.25	(2.04 ± 0.35) × 10⁻²	0.99
CH/EtOH 4d	0.17	0.17	0.18	(3.00 ± 1.5) × 10⁻⁴	0.99	0.22	(4.81 ± 1.42) × 10⁻²	0.99
Diluted dye								
CH/EtOH 1d	0.017	0.017	0.016	(4.50 ± 0.2) × 10⁻³	0.99	0.021	(25.75 ± 7.23) × 10⁻²	0.99
CH/EtOH 4d	0.017	0.011	0.011	5.00 ± 0.02) × 10⁻³	0.99	0.013	(58.36 ± 25.32)	0.96
DB/CH std								
Connected dye								
CH pH12	0.18	0.16	0.15	60.00 ± 0.09) × 10⁻³	0.99	0.19	(46.03 ± 8.96)	1
CH/NaOH	0.17	0.17	0.15	12.00 ± 0.02) × 10⁻²	0.99	0.20	(81.15 ± 20.11) × 10⁻³	1

Film	$q_{e,th}$	$q_{e,exp}$	$q_{e,calc}$	K	R^2	$q_{e,calc}$	K	R^2
CH/EtOH 1d	0.17	0.14	0.15	$(8.00 \pm 0.10) \times 10^{-3}$	0.99	0.21	$(27.56 \pm 6.42) \times 10^{-3}$	0.99
CH/EtOH 4d	0.17	0.16	0.18	$(8.00 \pm 0.1) \times 10^{-3}$	0.99	0.41	$(4.90 \pm 1.64) \times 10^{-3}$	0.85
Diluted dye								
CH pH12	0.018	0.019	0.019	$(2.00 \pm 0.04) \times 10^{-2}$	0.99	0.025	$(61.84 \pm 13.94) \times 10^{-2}$	1
CH/NaOH	0.017	0.018	0.021	$(2.00 \pm 0.04) \times 10^{-2}$	0.99	0.026	$(34.60 \pm 13.96) \times 10^{-2}$	0.98
CH/EtOH 1d	0.017	0.0056	0.004	$(9.00 \pm 0.50) \times 10^{-3}$	0.97	0.006	4.01 ± 1.93	0.99
CH/EtOH 4d	0.017	0.014	0.018	$(2.00 \pm 0.10) \times 10^{-2}$	0.98	0.3	7.69 ± 2.96	0.92

Table 3: Comparison of the pseudo-first-order, pseudo-second-order adsorption rate constants and calculated and experimental q_e values obtained at different Direct Red 83:1 concentrations

Film	$q_{e,th}$ (mg/cm²)	$q_{e,exp}$ (mg/cm²)	Pseudo-first order			Pseudo-second order		
			$q_{e,calc}$ (mg/cm²)	K (min⁻¹)	R^2	$q_{e,calc}$ (mg/cm²)	K (min⁻¹)	R^2
DR/CH new								
Connected dye								
CH	0.15	0.046	0.030	$(7.46 \pm 3.65) \times 10^{-4}$	0.36	/	/	/
CH/EtOH 1d	0.16	0.025	0.019	$(5.58 \pm 1.99) \times 10^{-4}$	0.48	0.007	17.36 ± 16.29	0.99
CH/EtOH 4d	0.16	0.018	0.014	$(23.95 \pm 3.13) \times 10^{-4}$	0.85	/	/	/
Diluted dye								
CH	0.016	0.010	0.011	$(133 \pm 10) \times 10^{-4}$	0.99	0.013	0.90 ± 0.39	0.98
CH/EtOH 1d	0.016	0.016	0.015	$(16.31 \pm 0.74) \times 10^{-4}$	0.97	0.009	1.06 ± 0.45	0.99
CH/EtOH 4d	0.017	0.016	0.015	$(5.00 \pm 0.02) \times 10^{-3}$	0.97	0.010	0.22 ± 0.74	0.99
DR/CH std								

Connected dye								
CH pH12	0.15	0.081	0.081	$7.66 \pm 0.28) \times 10^{-4}$	0.98	0.04	$(55.65 \pm 51.85) \times 10^{-2}$	0.95
CH/NaOH	0.15	0.017	0.012	$21.22 \pm 4.99) \times 10^{-4}$	0.64	0.01	13.98 ± 22.21	0.98
CH/EtOH 1d	0.15	0.043	0.029	$1.26 \pm 0.11) \times 10^{-2}$	0.94	0.05	$61.09 \pm 33.03) \times 10^{-2}$	0.99
CH/EtOH 4d	0.16	0.028	0.011	$10.02 \pm 2.23) \times 10^{-4}$	0.70	0.013	/	0.99
Diluted dye								
CH pH12	0.016	0.014	0.014	$28.32 \pm 0.81) \times 10^{-4}$	0.98	0.023	$8.27 \pm 3.46) \times 10^{-2}$	0.70
CH/NaOH	0.016	0.017	0.016	$20.12 \pm 0.82) \times 10^{-4}$	0.98	0.012	$54.86 \pm 18.57) \times 10^{-2}$	0.98
CH/EtOH 1d	0.016	0.012	0.113	$21.72 \pm 2.44) \times 10^{-5}$	0.86	0.01	2.31 ± 0.75	0.99
CH/EtOH 4d	0.016	0.011	0.007	$45.60 \pm 9.04) \times 10^{-4}$	0.70	0.009	5.24 ± 5.25	0.97

where $q_{e,th}$ (mg/cm²) is the amount of adsorbed dye at the equilibrium, C_0 is the initial dye concentration spectro photo metrically evaluated, A_{CH} is the geometrical area of the used CH film in cm² and V (L) is the dye solution volume. This equation derives from the application of Equation (2) to the calculation of the q_e values hypothesizing that all the dye has been adsorbed onto the CH film, whereby the amount of dye present in solution at the equilibrium is null ($C_t = C_e \approx 0$ M). More the calculated $q_{e,th}$ is closed to $q_{e,exp}$, the greater is the efficiency of various typologies of analyzed CH film. Comparing the obtained $q_{e,th}$ values for DY and DB with the $q_{e,exp}$ (shown in Table 1 andTable 2), an excellent dye uptake for each employed dye concentration relative to different film treatment is suggested.

However, as already observed by the analysis of the decoloration rate (Figure 2), not all CH film treatments improve the adsorption efficiency. For example there is a very high difference between the $q_{e,th}$ and the $q_{e,exp}$ in the case of the untreated CH film prepared with the modified procedure respect to the adsorption of DB, indicating that this new typology of CH film, if it is not dehydrated by ethanolic treatment, presents a low affinity for that dye. The obtained kinetic constant values suggest the effect of mass transfer, the initial dye concentration and affinity of DB respect to DY towards CH films. Figure 5(b) and Figure 5(c) show the linear form of the pseudo-first to the DR/CH film system. Only for diluted dye, the obtained q_e (Table 3) suggest an excellent dye uptake and kinetic constant values (Table 3) result to be in agreement with the q_t values. Calculated $q_{e,calc}$ values are agree with the experimental $q_{e,exp}$ values, also in this case.

Concerning the obtained results, probably an explication of this behavior could be found considering the concept of the point of zero charge (pH_{pzc}), indicative of the surface adsorption ability and of the type of surface active centers, applied to CH film. In particular the point of zero charge (pzc) is the pH value at which the surfacecharge is zero and it is typically used to quantify or define the electrokinetic properties of a surface. Consequently due to the presence of functional groups on the surface of the absorbent material, cationic or anionic dye will be absorbed at $pH > pH_{pzc}$ or $pH < pH_{pzc}$ [19]

The reported pH_{pzc} value for chitosan is 6.6 [38] so that the adsorption of a cationic dye may be favored at $pH > pH_{pzc}$, whereas anionic dye adsorption would be favored at $pH < pH_{pzc}$ where the CH surface becomes positively charged [18] . In our case the maximum dye uptake has been observed for all examined textile dyes when CH films have been previously dehydrated with EtOH, when treated in alkaline medium and when neutralized. However the amino groups of chitosan are protonated under acidic conditions, whereas in alkaline medium and/or in ethanolic solution, they result mainly unprotonated or partially protonated [26] . All employed dye are negatively charged and in an alkaline medium their charge density increases, further the parallel effect of EtOH is the increase of the hydrophobic character of CH films. This suggests that, in these dye/CH film systems, the electrostatic interactions are very weak and the adsorption processes are controlled mainly by physical forces such as Van der Waals forces, hydrogen bonds, polarity, dipole-dipole bonds, ϖ − ϖ interaction, etc [18] . In fact the ethanolic treatment of CH film induces a reorganization of CH chains within the film in a more favorable thermodynamic conformation since CH is insoluble in EtOH. In particular ethanol treated CH films result partially protonated, containing anhydrous "annealed" polymorph of chitosan and "type II" crystals of chitosan acetate, due to the simultaneous, but incomplete removal of the residual acid and water molecules from the films by immersion in ethanol [26] . The CH chain reorganization induced by EtOH favors the formation of hydrophobic domainsin chitosan [39] . This behavior leads to think that dye adsorption process would be favored in acidic medium where CH is completely protonated, but probably the excessive swelling of CH films in this case determine too weak CH film/dye interactions. It was not excluded the role of the lower solubility of dye in this condition. The worst results have been obtained in the case of DR, where the presence of OH moieties in the dye structure obstacle hydrophobic interactions between dye and CH films, especially if dye is present at high concentration. The CH film saturation in this case is achieved faster than for DY and DB. The effect of EtOH ensures the presence of hydrophobic interaction between anthracene and benzene moieties, and polar effect OH groups working mainly in opposition. More detailed analysis about the interactions CH/dyes by means of FTIR- ATR, morphology and mechanism of desorption and recycle are in progress. Also elucidation of EtOH and NaOH effects will be considered.

CONCLUSIONS

In this study, the adsorption of three anionic dyes, Direct Red 83:1, Direct Yellow 86 and Direct Blue 78, onto different types of CH films has been investigated focusing on equilibrium and kinetic process. The adsorption experiments confirm that CH modified films are effective for the adsorption of basic dye from aqueous solution. The kinetic studies show that the adsorption onto the different typologies CH follows a first order kinetics and Langergren model has been successfully applied. The adsorption of dyes onto CH film results to be dependent on various parameters such as aqueous solution pH, contact time and initial dye concentration. In particular, for CH std type films, good results have been obtained when the film is neutralized with NaOH (it is immersed in alkaline dye solution), and when it is treated previously with an ethanolic solution for 4 days. However changing the method of CH film preparation there is a remarkable improvement, especially when the film is treated with EtOH. The new method developed for preparing neutral CH films allows avoiding many washing cycles reducing the total cost of the process. Regarding the dye/CH film interaction DY and DB appear much more akin to CH films than DR, probably due to their ability in establishing hydrophobic interactions, which are prevented in the case of DR due to the presence of various OH groups.

ACKNOWLEDGEMENTS

This study was supported by the European "DYES4EVER" (Demonstration of cyclodextrin techniques in treatment of waste water in textile industry to recover and reuse textile dyes, LIFE12 ENV/ES/000309) within the LIFE+2012 program "Environment Policy and Governance project application". We gratefully acknowledge the skilful and excellent technical assistance of Mr. Sergio Nuzzo.

REFERENCES

1. Savenije, H.H.G. (2002) Why Water Is Not an Ordinary Economic Good, or Why the Girl Is Special. Physics and Chemistry of the Earth, 27, 741-744.http://dx.doi.org/10.1016/S1474-7065(02)00060-8.

2. Van der Oost, R., Beyer, J. and Vermeulen, N.P.E. (2003) Fish Bioaccumulation and Biomarkers in Environmental Risk Assessment: A Review. Environmental Toxicology and Pharmacology, 13, 57-149. http://dx.doi.org/10.1016/S1382-6689(02)00126-6.

3. Robinson, T., McMullan, G., Marchant, R. and Nigam, P. (2001) Remediation of Dyes in Textile Effluent: A Critical Review on Current Treatment Technologies with a Proposed Alternative. Bioresource Technology, 77, 247-255. http://dx.doi.org/10.1016/S0960-8524(00)00080-8.

4. Mirmohseni, A., SeyedDorraji, M.S., Figoli, A. and Tasselli, F. (2012) Chitosan Hollow Fibers as Effective Biosorbent toward Dye: Preparation and Modeling. Bioresource Technology, 121, 212-220. http://dx.doi.org/10.1016/j.biortech.2012.06.067.

5. Crini,G. (2005) Recent Developments in Polysaccharide-Based Materials Used as Adsorbents in Wastewater Treatment. Progress in Polymer Science, 30, 38-70.http://dx.doi.org/10.1016/j.progpolymsci.2004.11.002.

6.   Goncharuk, V.V., Kucheruk, D.D., Kochkodan, V.M., and Badekha, V.P. (2002) Removal of Organic Substances from Aqueous Solutions by Reagent Enhanced Reverse Osmosis. Desalination, 143, 45-51. http://dx.doi.org/10.1016/S0011-9164(02)00220-5.

7.   Lopez, F.A., Martin, M.I., Pèrez, C., Lopez-Delgado, A. and Alguacil, F.J. (2003) Removal of Copper Ions from Aqueous Solutions by a Steel-Making By-Product. Water Research, 37, 3883-3890. http://dx.doi.org/10.1016/S0043-1354(03)00287-2.

8.   Aksu, Z. (2005) Application of Biosorption for the Removal of Organic Pollutants: A Review. Process Biochemistry, 40, 997-1026. http://dx.doi.org/10.1016/j.procbio.2004.04.008.

9.   Park, D., Yun, Y. and Park, J.M. (2012) The Past, Present, and Future Trends of Biosorption. Biotechnology and Bioprocess Engineering, 15, 86-102. http://dx.doi.org/10.1007/s12257-009-0199-4.

10. Reddy, D.H.K. and Lee, S.-M. (2013) Application of Magnetic Chitosan Composites for the Removal of Toxic Metal and Dyes from Aqueous Solutions. Advances in Colloid and Interface Science, 201-202, 68-93.

11. Rabea, E.I., Badawy, M.E.T., Stevens, C.V., Smagghe, G. and Steurbaut, W. (2003) Chitosan as Antimicrobial Agent: Applications and Mode of Action. Biomacromolecules, 4, 1457-1465. http://dx.doi.org/10.1021/bm034130m.

12. Moczek, L. and Nowakowska, M. (2007) Novel Water-Soluble Photosensitizers from Chitosan. Biomacromolecules, 8, 433-438. http://dx.doi.org/10.1021/bm060454+.

13. Haldorai, Y. and Shim, J.J. (2014) An Efficient Removal of Methyl Orange Dye from Aqueous Solution by Adsorption onto Chitosan/MgO Composite: A Novel Reusable Adsorbent. Applied Surface Science, 292, 447-453. http://dx.doi.org/10.1016/j.apsusc.2013.11.158.

14. Crini, G. and Badot, P.M. (2008) Application of Chitosan, a Natural Aminopolysaccharide, for Dye Removal from Aqueous Solutions by Adsorption Processes Using Batch Studies: A Review of Recent Literature. Progress in Polymer Science, 33, 399-447. http://dx.doi.org/10.1016/j.progpolymsci.2007.11.001.

15. Jiang, X., Sun, Y., Liu, L., Wang, S. and Tian, X. (2014) Adsorption of C.I. Reactive Blue 19 from Aqueous Solutions by Porous

Particles of the Grafted Chitosan. Chemical Engineering Journal, 235, 151-157. http://dx.doi.org/10.1016/j.cej.2013.09.001.

16. Nandi, B., Goswami, A. and Purkait, M. (2009) Removal of Cationic Dyes from Aqueous Solutions by Kaolin: Kinetic and Equilibrium Studies. Applied Clay Science, 42, 583-590.http://dx.doi.org/10.1016/j.clay.2008.03.015.

17. Bulut, Y. and Aydın, H. (2006) A Kinetics and Thermodynamics Study of Methylene Blue Adsorption on Wheat Shells. Desalination, 194, 259-267.http://dx.doi.org/10.1016/j.desal.2005.10.032.

18. Yagub, M.T., Tushar Kanti, S., Afroze, S. and Ang, H.M. (2014) Dye and Its Removal from Aqueous Solution by Adsorption: A Review. Advances in Colloid and Interface Science, 209, 172-184. http://dx.doi.org/10.1016/j.cis.2014.04.002.

19. Gupta, V. (2009) Application of Low-Cost Adsorbents for Dye Removal—A Review. Journal of Environmental Management, 90, 2313-2342.http://dx.doi.org/10.1016/j.jenvman.2008.11.017.

20. Clarke, E. and Anliker, R. (1980) Organic Dyes and Pigments. The Handbook of Environmental Chemistry, 3, 181-215.

21. Pinheiro, H.M., Touraud, E. and Thomas, O. (2004) Aromatic Amines from Azo Dye Reduction: Status Review with Emphasis on Direct UV Spectrophotometric Detection in Textile Industry Wastewaters. Dyes and Pigments, 61, 121- 139.http://dx.doi.org/10.1016/j.dyepig.2003.10.009.

22. Robinson, T., McMullan, G., Marchant, R. and Nigam, P. (2001) Remediation of Dyes in Textile Effluent: A Critical Review on Current Treatment Technologies with a Proposed Alternative. Bioresource Technology, 77, 247-255. http://dx.doi.org/10.1016/S0960-8524(00)00080-8.

23. Verma, Y. (2001) Acute Toxicity Assessment of Textile Dyes and Textile and Dye Industrial Effluents Using Daphnia Magna Bioassay. Toxicology and Industrial Health, 24, 491-500. http://dx.doi.org/10.1177/0748233708095769.

24. Krajewska, B., Leszko, M. and Zaborska, W. (1990) Urease Immobilized on Chitosan Membrane: Preparation and Properties. Journal of Chemical Technology and Biotechnology, 48, 337-350. http://dx.doi.org/10.1002/jctb.280480309.

25. Krajewska, B. (1991) Chitin and Its Derivatives as Supports for the Immobilization of Enzymes. Acta Biotechnologica, 11, 269-277. http://dx.doi.org/10.1002/abio.370110319.

26. Fang, R., Cheng, X.S., Fu, J. and Zheng, Z.B. (2009) Research on the Graft Copolymerization of EH-Lignin with Acrylamide. Natural Science, 1, 17-22.http://dx.doi.org/10.4236/ns.2009.11004.

27. Khataee, A.R., Pons, M.S. and Zahra, O. (2009) Photocatalytic Degradation of Three Azo Dyes Using Immobilized TiO_2 Nanoparticles on Glass Plates Activated by UV Light Irradiation: Influence of Dye Molecular Structure. Journal of Hazardous Materials, 168, 451-457. http://dx.doi.org/10.1016/j.jhazmat.2009.02.052.

28. Jin, J., Li, L.S., Zhang, Y.J., Tian, Y.Q., Jiang, S., Zhao, Y., Bai, Y. and Li, T.J. (1998) Characterization and Structure of Side-On Azo Copolymers in Langmuir-Blodgett Films. Langmuir, 14, 5231-5236. http://dx.doi.org/10.1021/la980056v.

29. Karukstis, K.K., Perelman, L.A. and Wong, W.K. (2002) Spectroscopic Characterization of Azo Dye Aggregation on Dendrimer Surfaces. Langmuir, 18, 10363-10371.http://dx.doi.org/10.1021/la020558f.

30. He, Q., Ao, Q., Gong, Y.D. and Zhang, X.F. (2011) Preparation of Chitosan Films Using Different Neutralizing Solutions to Improve Cell Compatibility. Journal of Materials Science: Materials in Medicine, 22, 2791-2802. http://dx.doi.org/10.1007/s10856-011-4444-y.

31. Gibbs, G., Tobin, J.M. and Guibal, E. (2003) Sorption of Acid Green 25 on Chitosan: Influence of Experimental Parameters on Uptake Kinetics and Sorption Isotherms. Journal of Applied Polymer Science, 90, 1073-1080. http://dx.doi.org/10.1002/app.12761.

32. Cestari, A.R., Vieira, E.F.S., Pinto, A.A. and Lopes, E.C.N. (2005) Multistep Adsorption of Anionic Dyes on Silica/Chitosan Hybrid: 1. Comparative Kinetic Data from Liquid- and Solid-Phase Models. Journal of Colloid and Interface Science, 292, 363-372. http://dx.doi.org/10.1016/j.jcis.2005.05.096.

33. Cestari, A.R., Vieira, E.F.S., Dos Santos, A.G.P., Mota, J.A. and De Almeida, V.P. (2004) Adsorption of Anionic Dyes on Chitosan Beads. 1. The Influence of the Chemical Structures of Dyes and Temperature on the Adsorption Kinetics. Journal of Colloid and Interface Science, 280, 380-386. http://dx.doi.org/10.1016/j.jcis.2004.08.007.

34. Maghami, G.G. and Roberts, G.A. (1988) Studies on the Interaction of Anionic Dyes on Chitosan. Macromolecular Chemistry and Physics, 189, 239-243.

35. Shouman, M.A., Khedr, S.A. and Attia, A.A. (2012) Basic Dye Adsorption on Low Cost Biopolymer: Kinetic and Equilibrium Studies. IOSR Journal of Applied Chemistry, 2, 27-36.

36. Putnis, A. (1995) Introduction to Mineral Sciences. Cambridge University Press, Melbourne.

37. Chiou, M.S., Ho, P.Y. and Li, H.Y. (2004) Adsorption of Anionic Dyes in Acid Solutions Using Chemically Cross-Linked Chitosan Beads. Dyes and Pigments, 60, 69-84. http://dx.doi.org/10.1016/S0143-7208(03)00140-2.

38. Ali, H. (2010) Biodegradation of Synthetic Dyes—A Review. Water, Air, & Soil Pollution, 213, 251-273. http://dx.doi.org/10.1007/s11270-010-0382-4.

39. Philippova, O.E., Volkov, E.G., Sitnikova, N.L. and Khokhlov, A.R. (2001) Two Types of Hydrophobic Aggregates in Aqueous Solutions of Chitosan and Its Hydrophobic Derivative. Biomacromolecules, 2, 483-490. http://dx.doi.org/10.1021/bm005649a.

Chapter 7

Analysis of Pitting Corrosion on Steel under Insulation in Marine Environments

Susan Caines[a], Faisal Khan[a, b], and John Shirokoff [a]

[a]Safety and Risk Engineering Group, Faculty of Engineering and Applied Science, Memorial University of Newfoundland, St. John's, NL, Canada A1B 3X5

[b]Australian Maritime College, University of Tasmania, Launceston, TAS 7450, Australia

ABSTRACT

Corrosion under insulation (CUI) is an important issue in marine environments. Pitting corrosion is a significant contributor to this issue.

The ability to understand and model pitting behavior is integral to designing and maintaining assets in marine environments to decreased costs and increase safety and productivity. This paper reviews and analyses six categories of pitting knowledge to assess the current depth and breadth of understanding and to identify knowledge gaps in each category. The categories investigated are: identification of pitting, experimental methods, mechanism of pitting, modeling of pitting corrosion rates, remaining life assessment modeling, and risk-based inspections. This analysis finds that the depth of knowledge on pitting corrosion rate modeling and pitting mechanism is limited and requires further detailed study. The outcome of such study will strengthen pitting corrosion rate modeling, the accuracy of fitness-for-service assessments and risk-based inspection strategies.

INTRODUCTION

Corrosion under insulation (CUI) is a serious issue in marine environments. This type of damage can have catastrophic effects on production losses, health and safety, and the environment in the offshore industry if it is not identified before it degrades to a level where containment is threatened. CUI can take many forms in marine environments including pitting, uniform corrosion, and stress corrosion cracking. CUI can occur when moisture penetrates the insulation and helps to create a corrosion cell. This can occur in many ways including insulation damage or wicking, atmospheric wetness, or poor installation. If the component has a protective coating, breaks or holidays in the protective layer are also needed to expose the underlying metal to moisture. Pitting corrosion is thought to be the most common type of localized corrosion (Roberge, 2008) and is the focus of this review.

Pitting is a form of corrosion observed in some metals where corrosion is localized to small areas of degradation. It can lead to catastrophic consequences in marine applications. Small pits can progress through wall thickness and lead to a lack of containment of process materials or act as initiation site for stress corrosion cracks that can also lead to lack of containment. Brittle fracture of components is an issue. If pitting develops such that the strength of the member is affected, brittle failure can occur. This type of failure can be catastrophic and lead to a complete lack of containment or structural integrity of components.

The ability to predict pitting behavior is key to designing and maintaining assets in marine environments. If realistic models are not available, conservative corrosion rates are used and can lead to increased costs and decreased productivity.

Numerous studies and scholarly works have been done on the pitting behavior in steel for marine applications. The work varies from understanding how to identify pitting to predicting the likelihood of pitting and how it affects the service life of components. Studies on failure under insulation due to pitting in petrochemical applications have been conducted (Suresh Kumar, Sujata, Venkataswamy, & Bhaumik, 2008); however, no specific information on pitting under insulation in marine environments was found. Available literature was reviewed to determine the current state of understanding of pitting corrosion.

When reviewing the current understanding of pitting corrosion, the study was divided into six categories and the depth and breadth of available work analyzed to identify knowledge gaps in each category. The categories are:

1. Identification of pitting
    *   Inspection techniques
    *   Non-destructive evaluation
2. Experimental methods
    *   Simulating pitting behavior
3. Mechanism of pitting
    *   Phases of pitting
    *   Causes of pitting
    *   Factors effecting pitting potential
4. Modeling of pitting corrosion rates
    *   Models to predict corrosion rate
5. Remaining life assessment model

6.   Risk-based inspection
    • Predicting inspection/maintenance based on severity and susceptibility to pitting corrosion

Rating System for Depth and Breadth of Pitting Knowledge

A qualitative rating system was developed to characterize the literature in terms of understanding and application of monitoring, predicting, preventing, and controlling pitting in marine environments. Table 1 below illustrates the rating system.

Table 1: Rating system for depth and breadth of knowledge

Score	Nil = 0	Low = 1-3	Mid = 4-6	High = 7-9	Complete = 10
Depth	No understanding of topic	High level (shallow) understanding of topic. General concept is understood	Topic is understood. Competing theories by subject matter experts	Consensus between subject matter experts on topic	Complete understanding of topic
Breadth	No demonstration of broad application of theory	Limited application across fields	Increasing application of knowledge across environments/ industries	Demonstrated application across environments/industries	Complete demonstration of broad application of theory

The six designated categories of pitting in marine environments are summarized in the following sections. Each section includes an overview of current understanding and the depth and breadth of knowledge is identified.

CATEGORIZED REVIEW AND ANALYSIS

The six categories of pitting are reviewed, analyzed, and summarized in the following sections.

Identification of Pitting

The first step in understanding pitting in steel is to correctly identify the phenomenon. Pitting corrosion is characterized by small blemishes in the surface of a material.

Figure 1 illustrates pits on stainless steel in a simulated marine environment.

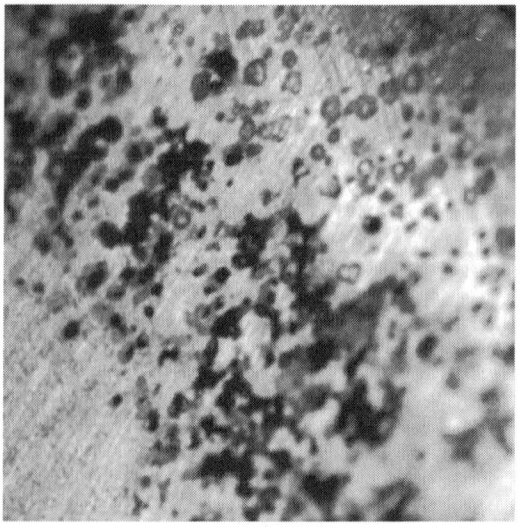

Figure 1: Pits on 304 stainless steel after exposure to simulated marine environment (3.5 g of NaCl per l H_2O).

Pits can form in many different shapes and sizes. Figure 2 shows some typical cross-sections of pits. The danger in pitting is that the size of the pit opening at the surface is not always indicative of the amount of sub-surface corrosion. This can lead to structural instabilities in components that may appear to have little surface damage.

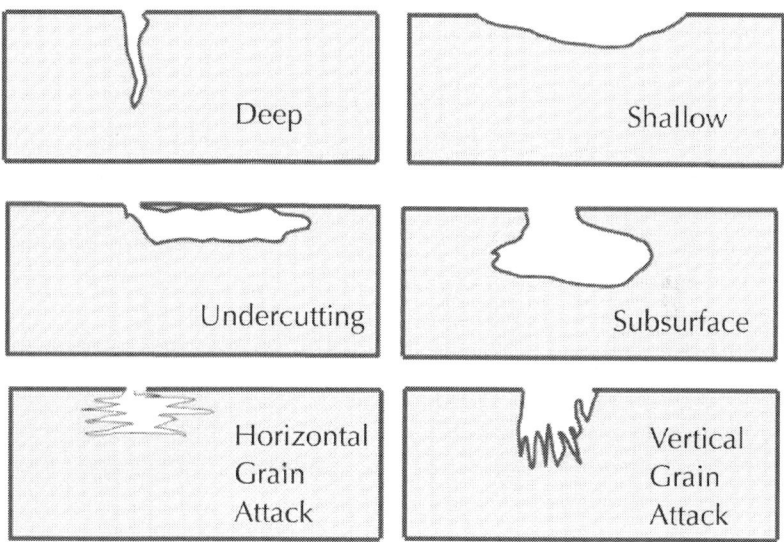

Figure 2: Typical cross-sectional shapes of corrosion pits (Phull, 2003 and Roberge, 2008).

While shallow pits are easier to examine and are unlikely to affect the structural integrity of the component, they can act as stress concentrators and initiate stress corrosion cracking (SCC). SCC is another corrosion mechanism seen under insulation in marine environments. Its contribution to CUI is also explored in this work.

There are many techniques that can identify the presence of pitting. This part of pitting corrosion is well understood and well documented by Davies and Scott, 2003, McIntyre and Vogelsang, 2009 and Phull, 2003, and Roberge (2007). The main techniques identified in the American Society for Metals (ASM) Handbook (Phull, 2003) to identify pitting are as follows:

- Visual inspection
- Metallographic examination

- Mass loss
- Pit depth measurement
- Non-destructive inspection

Visual Inspection

The American Society for Testing and Materials (ASTM G46, 2005) Standard G46 "Standard Guide for Examination and Evaluation of Pitting Corrosion" describes visual inspections as inspection that can be done in ambient light to determine location and severity of pitting. Pictures are often used to document the difference in appearance of pits before and after removal of corrosion products. This technique is the easiest to employ, requires no specialized equipment and is relatively inexpensive. More detailed descriptions of visual inspections are well documented in Byars, 1999, Heidersbach, 2011 and Roberge, 2007, and Visual inspection (1989).

More complex visual inspection techniques are used to evaluate areas that are difficult or dangerous for personnel to access. These visual inspections are facilitated through use of video and robotics; both remotely operated and autonomous.

Remote operated vehicles (ROVs) attempt to replace human visual inspections to increase safety, reduce cost, and increase efficiency (Terribile, Schiavon, Rossi, Zampato, & Indrigo, 2007). These vehicles use video to allow inspectors to guide and inspect areas that are difficult to reach and/or are dangerous. ROVs can detect external corrosion, damage, and anode wear in deep water pipelines (Kros, 2011)

Work is being conducted to adapt Autonomous Underwater Vehicles (AUVs) to perform visual inspections in underwater structures and pipelines (Courbot et al., 2013, Mcleod et al., 2012 and Yu and Ura, 2002). These inspections would include high resolution photographs of the length of the pipe, real-time image processing, and location tagging for future inspection.

Metallographic Examination

Metallographic examination is an investigative technique that can be used to determine the size, shape, and density of corrosion pits. It is one of the most important examination techniques as it can yield quantitate information on pitting corrosion. This technique is also used to verify true pits versus metal dropout from other corrosion mechanisms or to investigate corrosion rate correlation to inclusions and microstructure (ASTM G46, 2005). Figure 3 shows a cross-section of a corrosion pit on 316 L Stainless steel. Measurements for maximum pit depth (C) and pit width (B) may be recorded as indicated.

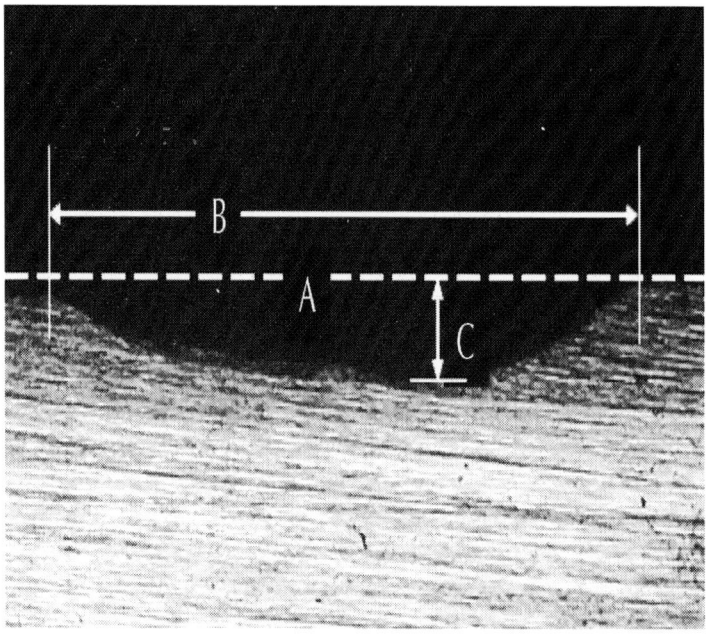

Figure 3: Cross-section of pitting corrosion on 316 L stainless steel. A: Original surface, B: Pit width, C: Pit depth (Snow & Shirokoff, 2008).

Metallographic examination is typically a destructive analysis technique as the specimen must be cut from the component and examined with a microscope. In-situ metallography can be used when removal of the component is not feasible. This type of metallography uses surface replication and does not produce the same quality micrographs as traditional destructive techniques (Jana, 1995).

Simultaneous in-situ optical and electrochemical methods for identifying and measuring pitting corrosion have also been reviewed and measured (Power and Shirokoff, 2012 and Power and Shirokoff, in 2013). This approach was used to measure corrosion in 316 L stainless steel subject to simulated seawater conditions and in industrial sulfuric acid environment simulating hydrometallurgical recovery of metals from nickel sulfide ores. Power and Shirokoff found that this innovative approach to studying the surface microstructural changes in real time could successfully correlate to the electrochemical response at the surface under aerated and deaerated conditions. The technique is a low cost and practical method to investigate samples under constant temperature conditions in a custom built corrosion cell attached to an electrochemical probe, potentiostat–galvanostat, optical microscope, digital camera, and VHS-VCR-DVD recording system.

Proper surface preparation of samples is important; poor preparation can lead to inaccurate measurements, observation and even destruction of the sample. Sample preparation includes cutting a sample (sectioning), mounting of small samples if needed, cleaning of a surface, and polishing (ASTM E3, 2011 and ASTM G1-03, 2011). All of these steps must be conducted carefully and appropriately for the material and environmental conditions. Proper care must be taken to ensure the preparation methods do not affect the important surface.

Examples and details of metallography are widely discussed, a few important sources are ASTM E3, 2011, ASTM G1-03, 2011, Gale and Totemeier, 2004 and Reardon, 2011, and Vander Voort, 1999 and Vander Voort, 2004.

Mass Loss

Mass loss techniques are used to determine the amount of material lost due to corrosion. This is accomplished by a systematic measurement of the mass loss over a specific period of time.

The application of mass loss studies to pit evaluation is limited. Mass loss due to this type of localized corrosion can be too small to allow for identification through this method. Some standards for pitting evaluation include mass loss as a possible technique includes ASTM G46 (2005) and Phull (2003). This technique may be useful if pitting is the predominant corrosion mechanism and the amount of general corrosion in minimal. Other sources do not include mass loss as a viable technique for pitting identification (Baboian, 2005, Heidersbach, 2011 and Jones, 1996, pp. 200–213).

This method is most useful in evaluation of uniform corrosion, corrosion that affects the total surface area of a component. A standard methodology for preparing samples for mass loss evaluation such as ASTM G1-03 (2011), 'The Standard Practice for Preparing, Cleaning, and Evaluating Corrosion Test Specimens", are used.

Pit Depth Measurement

Pit depth measurement is a key technique in pit identification and evaluation (ASTM G46, 2005 and Phull, 2003). ASTM G46 (2005) describes different methods to evaluate pit depth. Metallography can be used to evaluate a vertically sectioned pit (ASTM G46, 2005). The depth of the pit can then be measured with a calibrated eyepiece. The limitation of this method is that the deepest pit may not be selected for evaluation. Machining is another method discussed by the standard. This method involves systematic machining of a pitted surface and subsequent thickness measurement to determine pit depth. This method can be used to find the maximum pit depth and to determine the number of pits with specific depths. These two methods are destructive and cannot be used in service. Alternatively, a depth gage may be used in service to determine pit depth. This method uses a calibrated depth inserted into a pit. This method is limited to pits that are large enough at the base to allow full penetration of the gage and that have not experienced undercutting or a direction change (ASTM G46, 2005).

Jasiczek, Kaczorowski, Kosieniak, and Innocenti (2012) have identified a new non-destructive method to evaluate pit depth using Confocal Laser Scanning Microscopy (CLSM) that has shown potential to further the ability to measure pit depth. CLSM creates a three dimensional image of a material surface (Clarke & Eberhardt, 2002, pp. 228–236) that can be analyzed to determine pit depth. The authors showed that this technique can reliably measure pit depth and had the potential to evaluate additional pit characteristics such as diameter and volume.

Non-destructive Testing

Non-destructive testing (NDT) is a key technique used in industry to evaluate the current state of components and equipment in service and to aid in maintenance planning. NDT is used to identify, monitor and qualify many types of issues in industry during operations and during short operational shut-downs. Removal of components from a working facility is not practical so NDT becomes more important for defect evaluation.

ASTM G46 (2005) Standard describes NDT applicable to identifying pitting corrosion. NDT is well-established; however, these techniques are not as effective at characterizing pitting as destructive methods. NDT also requires specialized training to ensure realistic results. Many references are available that describe different types and applications of NDT for pit identification (Heidersbach, 2011, Roberge, 2007, Roberge, 2011 and Shreir et al., 1994). A brief description of each NDT is presented below.

Radiography

Pitting is readily detected by radiography and this technique is routinely applied in service identify and monitor corrosion (Heidersbach, 2011). In this technique, radiation/X-rays passes through the component under investigation and the intensity of the exit rays indicates changes in thickness. To successfully identify pits, the depth must be larger than 0.5% of the metal thickness (ASTM G46, 2005). This technique can quickly identify corrosion issues however, only small areas are inspected at a time, the 2D image give no depth information and access to both sides of a component is required (Heidersbach, 2011).

When a component is insulated, the insulation has traditionally been removed for inspection and identification of pitting. This is a time consuming and costly operation. Pechacek (2003) introduced a new method for inspecting insulated vessels using profiler portable real-time radiography (PPRTR) (Pechacek, 2003). This technique can quickly identify areas of concern, both gradual loss, indicating general corrosion and abrupt wall thickness changes indicated localized (pitting) corrosion. This method allows for more thorough coverage of long insulated pipe lengths and can identify areas that require further NDT to determine action.

Electromagnetic

This type of evaluation technique includes eddy current, magnetic particle, and microwave techniques (Rao, Jayakumar, & Raj, 2007). These techniques are used on electrically conducting materials and use induced magnetic fields to detect defects (ASTM G46, 2005). The discontinuities in the material are identified by their effect on electrical conductivity or magnetic permeability or dielectric permittivity.

Sonics

This technique uses sound energy to find the size and location of pits.

Pellegrino and Nugent (2012) investigated remote visual inspection (RVI) with 3-D phase measurement to size pits in compressor blades and the characterization of pipe wall pitting with phase-array ultrasonic testing (PAUT) with dual transducer. They report that new 3-D phase measurement technology can measure pits with diameters as small as 0.1 mm and depths as shallow as 0.025 mm. They indicate that these measurements are accurate and can be conducted quickly. Traditional ultrasonic transducers have difficulty in accurately measuring and identifying pits. These limitations are due to limited inspection area. PAUT was developed to increase the accuracy through the use of multiple receiver elements. This increases the area inspected and the likelihood of identifying and measuring the deepest pits.

Jirarungsatian and Prateepasen (2010) in their paper discuss acoustic emission (AE) as a method for detecting both pitting and uniform corrosion. This in-service method detects transient waves from energy released from localized material sources to directly measure corrosion failure mechanisms. They indicate that ambient noise has been the main issue preventing field use of this detection method.

Penetrants

Penetrant examination is a non-destructive technique that identifies surface defects on a non-porous surface. Because pitting occurs at the surface, this method is widely used to find and classify pits (Borucki,

1989, pp. 71–88). This method detects surface pits through the application of a liquid penetrating material; the liquid must penetrate defects through capillary action over the dwell time (Raman, 2007). Intensity of the color and the rate of bleed out both indicate the size of the defect (ASTM E1417, 2013 and ASTM G46, 2005).

There are two types of penetrants, fluorescent and visible. The appropriate type of penetrant is chosen based on many factors including type of flaw detected, surface condition, and sensitivity required (Borucki, 1989). Fluorescent penetrant is more reliable and sensitive than visible penetrants and are used more often. Visual penetrants are usually red and must be viewed under white light while fluorescent penetrants are typically green and glow under ultraviolet light.

Electrochemical Identification

Another method to evaluate pitting corrosion is through the use of electrochemical methods. ASTM F746, 2009, ASTM G61-86, 2009 and ASTM G150-99, 2010 has several standards that describe these techniques. These techniques use applied current or potential to establish the relative performance of materials in an environment. These techniques are usually applied in laboratory testing. Use of these techniques for on-line monitoring is limited due to the high applied potentials that permanently effect components under evaluation (Phull, 2003).

One newer method of electrochemical monitoring of corrosion that has the potential to be used on-line for the detection of pitting corrosion is electrochemical noise (EN) measurement techniques. EN technique is a passive method of corrosion monitoring as no applied current is required (Frankel, 2008). This method measures deviation from the naturally occurring electrochemical potential (Reiner & Bavarian, 2007). This variation is due to corrosion and can be measured. This technique has shown good correlation in detecting the formation of localized corrosion (Estupiñán-López et al., 2011). A single electrode monitoring probe has been successfully used in applications where the structure can be used as a current return path (Eden & Kane, 2005). It is expected that this method can be used to indicate when corrosion is occurring; however the challenge with electrochemical monitoring is to directly determine corrosion rate in non-immersed applications (Klassen & Roberge, 2003).

Innovative Techniques

While there are many standards and related scholarly works describing well-established pitting identification methods, new and improved techniques are also being investigated.

Papavinasam, Attard, Doiron, Demoz, and Rahimi (2012) investigated five non-intrusive inspection techniques on test pipes with artificially implanted pits over 12 years. Their work established the reliability of each of the techniques based on a number of criteria. Table 2 summarizes their findings as follows.

Table 2: Summary of findings by Papavinasam et al. (2012)

Technique	Reliability	User-friendliness	Sensing element	Surface area/Sensor ratio.	Remote monitoring	Boundary and limitations	Conclusions
Ultrasonic-handheld	Can detect location of defect or pit. Reliable to error of ±0.25 mm	Low set-up time, Portable, applicable for hazardous field conditions	Piezoelectric crystal to mechanical energy. Determines thickness without accessing the internal pipe surface	Manual scanning	No remote monitoring, onsite data collection by technician	Requires: physical contact and couplant. Shape influences results. Results dependant on experience and skill.	Most ideal, current, non-intrusive technique
Ultrasonic-fixed	Cannot detect location of defect or pit (fixed location). Reliable to error of ±0.25 mm. Cannot be calibrated; long-term reliability not ensured	Longer set-up time. Can be used in hazardous field conditions. Not portable	Piezoelectric crystal to mechanical energy. Determines thickness without accessing the internal pipe surface	N/A. Fixed Location	Theoretically suitable for remote monitoring	Requires: physical contact and couplant. Measurements only at locations where sensors installed (2.5 cm). Results independent of experience and skill after installation.	Development of liquid couplant: Not dry over time and long-term adherence to substrate
Electrical probe	Reliability dependant on: number of pins, distance between pins, contact resistance, applied current, and accuracy of resistance measurement instrument	Two (2) options. 1. Permanently spot welded to structure: Low contact resistance, restricted use in some applications and jurisdictions. 2. Spot welded onto pipe section and clamped to structure. Portable with higher contact resistance	Based on Ohm's Law where resistance is inversely proportional to wall thickness. 2 pins to apply current, 2 pins to measure potential.	Dependant on number of pins. Increased distance between pins, measured area increases, sensitivity to wall loss decreases.	Appropriate for remote monitoring.	Difficulties due to defect geometry. Impractical for large surfaces. Relies heavily on operator skill and experience. No testing or evaluation by regulatory or by standards making body	Establishment of a relationship between geometry of pins, wall thickness, and resistance measurement needed

Hydrogen permeation	Indicates cannot be reliably used to measure pitting corrosion rates	Can be moved. No welding, machining or use of epoxies. Surface is not modified.	Measures pressure increase or hydrogen gas.		Not suitable for remote monitoring	Does not directly measure wall thickness. Not capable of detecting defect location. Provides a corrosion rate for a general area	Suitable applications required
Fiber-optic	No correlation established between measurements to physical measurement of pit depths	Easy attachment. Very fragile. Difficult to remove once attached.	Cable is both sensor and communicator. Microstrain of cable measured to determine wall thickness.	Area covered is proportional to length of cable	Suitable for remote monitoring	Very new and needs to be proven. Fiber very fragile. No operator training available	Fragile fibers limiting advancement

Holme and Lunder (2007) describe pit characterization using White Light Interferometry and software analysis. The program analyses low resolution images to locate and direct the white light interferometer to capture high resolution images that are analyzed to find depth, volume and maximum width of pits. This technique is limited to pits that do not experience undercutting and is best utilized to pit initiation and early propagation. This technique generates 3-D experimental data of pits.

Analysis of Pit Identification Knowledge

The effectiveness of these non-destructive evaluations is important to operations in harsh marine environments and needs to be understood. These techniques will be used in the design of components to select the best material, manufacturing, and installation practices, and in operations to plan inspection and maintenance scheduling, and in developing models for predicting asset lifecycles (Heerings, Trimborn, & den Herder, 2006).

The referenced work summarized above with respect to pit identification indicates that there is significant information on pit identification techniques available and that they are well understood. New techniques that improve accuracy and reduce human errors are currently in development and will lead to increased confidence in pit identification. The depth of understanding of pit identification can be considered at a ranking of 8. Techniques are available to quantify pitting depth and severity in many marine applications. Further study is ongoing to improve on-line monitoring of pitting corrosion and to further understand correlation between measured values and pit depth. The breadth understanding of pit identification is determined to be in the mid-range (6). While information is available to assess pitting from many different approaches including laboratory, field and on-line monitoring, there remain many instances where timely, cost effective pit identification techniques are unavailable.

Experimental Methods

The evaluation of pitting behavior is required to fully understand and predict the phenomenon. Determining relationships between many factors including composition, temperature, and environmental conditions are conducted through experiments. Pitting rates determined through experimental methods are generally used in prediction models. Because these rates are used to conduct remaining life assessments, experimental methods need to be conducted such that the results can be extrapolated over longer periods of time.

Many different methods can be used to collect the data required to further understand pitting corrosion. Information can be gained from in-service observation and experimentation, field testing, and laboratory experiments. For many situations, there are standard methods available however; much work has been conducted using generalized corrosion test planning methods that are specific to the situation under review. The method described by Cramer and Jones (2005) includes the general five step design:

- Goal and Objective definition
- Corrosion Test Design
- Protocol development
- Test Engineering
- Test Modification

These five general steps are used to adapt current standards to unique situation while ensuring the results are recorded, evaluated, and reported in a systematic, repeatable manner.

Standards

Standards are available to evaluate pitting susceptibility of various materials and environments (ASTM F746, 2009, ASTM G48-11, 2011, ASTM G61-86, 2009 and ASTM G150-99, 2010). These standards can be used as a comparative tool to determine the likelihood of pitting in specific circumstances and cannot indicate behavior of materials in service.

ASTM G48-11 (2011), "Standard Test Methods for Pitting and Crevice Corrosion Resistance of Stainless Steels and Related Alloys by Use of Ferric Chloride Solution", describes six test methods to determine relative pitting and crevice corrosion resistance of stainless steels. Methods A, C, and E of this standard deal specifically with pitting corrosion. Method A is a Ferric Chloride Pitting Test and C and E rank materials based on critical pitting temperature (CPT). For this standard,

CPT is the temperature at which pitting of a depth of at least 0.025 mm is expected. The results of these methods are used for comparison and ranking of materials in chloride environments. These tests are accelerated and the rate and extent of pitting are not representative of expected field results.

ASTM G150-99 (2010), the "Standard Test Method for Electrochemical Critical Pitting Temperature Testing of Stainless Steels", includes procedures for determining the potential independent critical pitting temperature (CPT) of stainless steels using electrochemical methods. For this test, CPT is found when the measured current rapidly increases. The onset of pitting above this CPT is visually verified after the test. Again, the standard procedure accelerates corrosion in a way that does not represent any actual service environment.

ASTM G61-86 (2009), "Standard Test Method for Conducting Cyclic Potentiodynamic Polarization Measurements for Localized Corrosion Susceptibility of Iron-, Nickel-, or Cobalt-Based Alloys", is used to determine the relative susceptibility of a material to pitting. This is recorded as the potential at which the anodic current increases rapidly. Higher potentials (more noble) are an indication of increased resistance to pitting. This procedure induces corrosion and the results are not intended to indicate the rate of pitting expected in service.

These methods are used to further the understanding of pitting behavior and to assist in determining the effects of change on the resistance of a material to pitting. Siow, Song, and Qiu (2001) used a method similar to ASTM G61 to evaluate the complex effect of alloying and microstructure on pitting. They found that the effect of alloying is complex and that the alloying elements may increase or decrease the effects of other alloying elements. They also report that pits started at the ferrite–austenite border and then spread into the austenite and ferrite phases.

It should be noted that there are standardized tests for accelerated corrosion including ASTM B117-11 (2011), "Standard Practice for Operating Salt Spray (Fog) Apparatus" and ASTM G85 (2011), "Standard Practice for Modified Salt Spray (Fog) Testing". Both of these standards allow for increased severity of a corrosive environment to accelerate corrosion. The results from these tests can indicate pitting however; there is limited correlation between field results and these accelerated tests (Acevedo-Hurtado et al., 2008).

Non-standard

Experimental methods can also be developed to simulate and evaluate pitting behavior in specific situations and to evaluate pitting resistance changes due to controlled factors. Researchers can adapt accepted standards to tailor methods to these situations.

Studies use accelerated testing to rank materials in terms of their pitting resistance rather than to determine corrosion rates (De-Abreu, Helander, Suarez, Manko, & Clark, 2012). ResearchersLothongkum, Vongbandit, and Nongluck (2006) and Moran, Frankel, & Kim (2011) have used modified cyclic potentiodynamic polarization to evaluate pitting corrosion resistance and Krakowiak and Darowicki (2002) developed their own methodology to determine the effect of temperature rate change on the critical pitting corrosion temperature. They used three electrode measurement vessels with controlled temperature change and determined that the CPT does not depend on temperature change rate in their experimental range.

Researchers from Kushiro National College of Technology and Kitami Institute of Technology in Japan developed an experimental procedure to study the effect of a freeze-thaw cycle on pitting of welded austenitic stainless steel (Takahashi, Shibano, Ishitsuka, & Kobayashi, 2012). This procedure was developed to help evaluate structures in coastal regions with severe chloride containing environments. The researchers incorporated the Japanese Industrial Standard (JIS G 0578, 2000) "Method of ferric chloride tests for stainless steels" and modified the test solution composition and temperature because the environment under investigation was not comparable to standardized tests. The study concluded that pitting corrosion was more severe in the freeze-thaw specimens than in the constant thaw specimens as determined by increased mass loss. Also, they determined that tensile residual stress is related to accelerated pitting corrosion.

Electrochemical impedance spectroscopy (EIS) is another method that uses applied current to study pitting corrosion. Sorg and Ladwein (2009) used this method to determine the susceptibility of a material to pitting corrosion in the presence of low conductivity electrolytes. They found that EIS allowed for polarization resistance analysis in low conductivity fluids (Sorg & Ladwein, 2009). Jia et al. (2011) used staircase EIS to evaluate pitting in 316 L stainless steel. They found that passive film breakdown was the most likely cause of pitting corrosion (Jia, Du, Li, Yi, & Li, 2011).

Field Testing

Field testing is an important method to gather long-term information about corrosion in a natural environment. To study pitting in real situations, field testing has been conducted by researchers (Chaves and Melchers, 2012, Melchers, 2004 and Phull, 2003). Field testing had helped to demonstrate that while short term testing and accelerated testing are valuable in understanding corrosion, they can be misleading in predicting pitting behavior over the long-term (Acevedo-Hurtado et al., 2008 and Chaves and Melchers, 2012).

Atmospheric tests are another important method of gathering information and evaluating pitting corrosion in marine environments. ASTM has many relevant standards that can be used in pitting corrosion field testing including ASTM B826, 2009 and ASTM G33, 2010, and ASTM G50 (2010). While these standards are not specific to pitting, if the mechanism of corrosion of the material being studied in the tested atmosphere is pitting, they can be used in pitting corrosion studies.

Analysis of Experimental Methods Knowledge

Experimental methods to compare pitting resistance of materials are well-established and can be successfully modified to accommodate different corrosive environments. This is useful in identifying likely candidates for service applications through relative resistance to a specific environment. Standard laboratory methods for pit evaluation have not been developed to attempt to determine corrosion rates of pitting that can be translated to real life situations. Accelerated corrosion testing is not valid to determine pitting rates in service. Field data has shown that short term testing cannot be relied on to predict long-term corrosion behavior. For these reasons, the experimental methods category is classified as a mid-range depth with a score of 4 and a wide breadth score of 8 due to the prolific application of testing across industries, material types, and corrosive environments.

Pitting Mechanism

One application of experimental methods is to aid in determining the mechanisms involved in pitting behavior. In marine applications, pitting usually occurs in coated or naturally protected materials. Corrosion resistance in stainless steel is partially due to a naturally occurring passive oxide layer that forms over the surface of the material. For other types of steel, such as carbon steel, corrosion protection is sometimes due to an applied protective coating. Although these protective layers prevent corrosion over the bulk of an asset, it is where the layer fails or is inconsistent that localized pitting corrosion can occur.

Pitting capitalizes on breaks in the protective layer. A breakdown in the protective layer, either natural or applied, provides a nucleation point for the formation of pits in the presence of an electrolyte containing an aggressive anion (Szklarska-Smialowska, 2005). For marine operations, this ion (Cl⁻) is readily available in seawater and marine atmospheres.

According to Schumacher (1979), some metals exposed to a corrosive environment will develop pits due to salt particles or other contaminants. Other factors that contribute to pitting including:

- Inclusions
- Discontinuities in protective coating (both natural and applied)
- Surface defects

The mechanism of pitting is not fully understood however most theories look at pitting as a combination of stages. Pitting corrosion damage is identified by Engelhardt and Macdonald (2004) as a three stage event including:

Stage 1: Nucleation: in this stage, pits are initiated (nucleated)

Stage 2: Propagation: here, some pits begin to grow

Stage 3: Repassivation: this stage includes pits that cease to continue to grow.

These stages can occur simultaneously leading to large variation in the location, depth, severity, and density of pitting. This contributes to the complexity of predicting pitting rates and to the current view of pitting corrosion as a random process.

Nucleation

The nucleation of pits is influenced by surface defects that may be due to manufacturing issues, installation problems, maintenance procedures, and/or environment changes (Baboian, 2005 and Heidersbach, 2011)

The sight of pit initiation (nucleation) can be caused by many different factors:

Damage to protective oxide layer (chemical or mechanical)

- Environmental factors causing protective layer breakdown
- Acidity, low dissolved oxygen
- High chlorine concentration
- Damage to applied protective coating
- Poor application of protective coating
- Material structure non uniformity

All of these factors lead to adjacent anode and cathode sites available for corrosion if an electrolyte is present.

Pit nucleation sites can be categorized in two different combinations (Roberge, 2008):

Combination 1: Abnormal anodic site surrounded by normal cathodic surface where the anodic sites will corrode.

Combination 2: Abnormal cathodic site surrounded by normal anodic surface where pitting corrosion will occur.

Figure 4 below illustrates these two combinations:

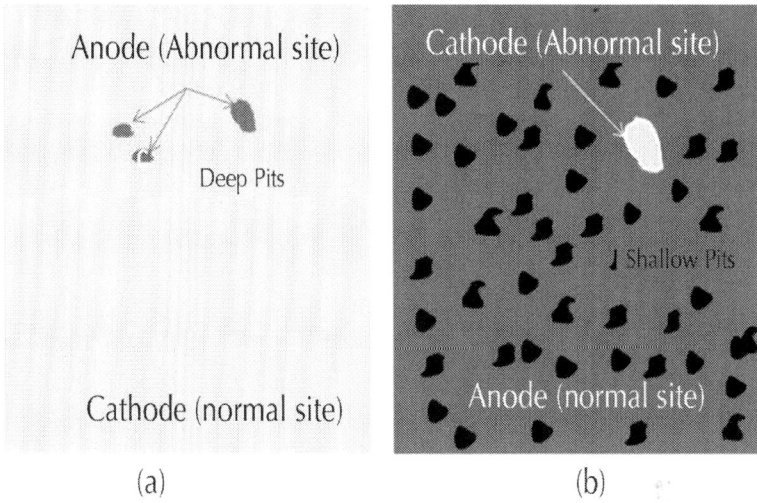

Figure 4: Pit nucleation site combination. A: Combination 1, abnormal anode, normal cathode. B: Combination 2, normal anode, abnormal cathode.

Combination 1 indicates a higher expected corrosion rate and more severe pits. This is expected due to the difference in surface area of the anode and cathode. The cathode (normal surface) has a much larger surface area than the anode and will corrode the smaller anode quickly and produce deeper pits. This is expected in materials with an applied coating.

Combination 2 can lead to more extensive overall pitting of the surface however this can be a benefit when the abnormal cathodic area is much smaller than the surrounding anodic (normal) surface as pits in this combination tend to be shallow and less likely to extend through the wall thickness. Pits can be seen over the bulk of the metal with only a small area of unaffected local cathode. Roberge (2008) considers this combination to be the most common.

Passive Film Breakdown

In stainless steel, the breakdown of the passive film provides the site for pit nucleation. These breakdown sites are susceptible to corrosion. Predicting this breakdown is difficult and no generally accepted model has been identified. Further discussion on modeling of pitting can be found in the sections that follow (Sections 2.4, 2.5 and 2.6).

198   Cost Estimating Manual for Pipelines and Marine Structures

Passive films are present on the surface of stainless steels in the presence of oxygen. At low temperatures, a true oxide layer is not formed but a thin passive film is formed and acts as a barrier and provides corrosion resistance (Grubb, DeBold, & Fritz, 2005, pp. 54–77). This film should be continuous, non-porous, insoluble, and self-healing to fully protect against corrosion. Alloying elements and environmental conditions determine the success of this protection.

In marine applications, hydroxide ions help to form the passive film and chloride ions attack the film, causing openings for pit formation. The tug of war between these reactions limits pit initiation (Novak, 2007, pp. 40, 45, 53, 54). If formation of the passive film is the stronger reaction, the opportunities for pit initiation are reduced; if the breakdown reaction dominates, pitting is encouraged.

E_{pit} is a generally accepted indication of resistance to pitting however there remains uncertainty due to experimental scatter, the dependence of E_{pit} on experimental parameters, and experimental evidence of pit initiation below E_{pit} (Frankel, 1998).

Electrochemical studies using cyclic anodic polarization indicate that pits form at a potential above a characteristic potential E_{pit} (Jones, 1996). This has been shown to be valid for both electrically and chemically induced potentials.

The susceptibility of metals and alloys to pitting corrosion can be estimated using polarization curves (Szklarska-Smialowska, 2005) and can be developed through standardized methods discussed in Section2.2.1 (ASTM G61-86, 2009). The curves are used to find pitting potential (E_{pit}) and repassivation potential (ER). A schematic can be seen in Figure 5.

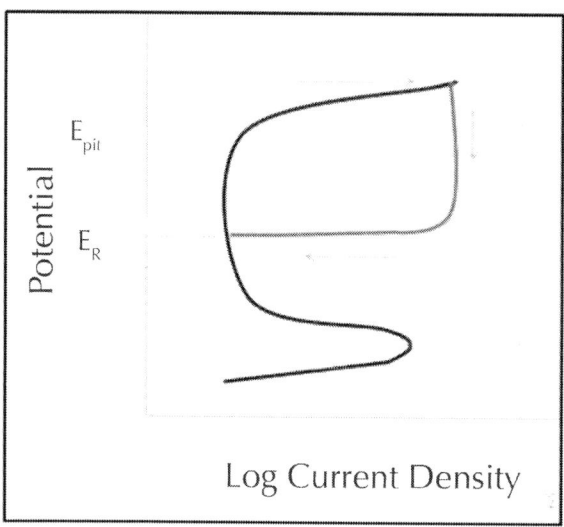

E_{pit}

E_R

Potential

Log Current Density

Figure 5: Schematic of anodic polarization curves for a metal immersed in a solution containing aggressive ions (Frankel, 2008, Jones, 1996 and Szklarska-Smialowska, 2005).

Higher (positive) E_{pit} for a material in a given environment indicates greater resistance to pitting (Jones, 1996 and Szklarska-Smialowska, 2005). If the potential is reduced below E_{pit}, the surface may repassivate and pit growth can stop. If the potential is between E_{pit} and *ER*, pitting is expected (Craig, 1991, pp. 109–126).

Propagation

This stage of pitting is where pits grow and have the potential to increase beyond wall thicknesses and lead to leaks.

For pits to propagate, certain conditions must be met:

- E_{pit} must be exceeded and remain above *ER*
- An aggressive ion must be present
- Localized breakdown of passive or applied film

Pits are thought to initiate when the potential of the cell exceeds the pitting potential (E_{pit}) of the material in a given environment and grow (propagate) if the potential remains above the repassivation potential, *ER* (Frankel, 1998).

There are many theories for the mechanism of pit growth. Jones (1996), in his text book "Principles and Prevention of Corrosion" describes pit growth as a autocatalytic process. Within a pit, Fe^{2+} ions attract negative ions (Cl^- in marine applications) and through hydrolysis creates a porous $Fe(OH)_2$ cap over the pit. This creates a self-propagating system where the increased acidity in the pit cavity increases corrosion of the steel walls of the pit. Cl^- ions migrate through the cap into the pit and Fe^{2+} migrates out.

Repassivation

Pits that continue to grow in stage 2 are the pits that will eventually threaten the integrity of an asset; however, all pits that are initiated (stage 1) and propagate (stage 2) do not always continue to grow. Pits can repassivate and stop growing. This is common in materials that have a naturally produced passive layer such as some stainless steels. In steels that are protected by an applied coating pitting may be stopped by reapplication of a coating. Repassivation can be thought to occur below the E_R.

Work by Novak (2007) suggests increased internal resistance of the local cell within the pit is the reason for pit death (repassivation). The author suggests that the increase in resistance may be due to:

- The pit filling with corrosion products
- Filming of the cathode that limits reaction.
- Drying out of the surface (if rewetted, pits may reinitiate and continue to grow)

Analysis of Pitting Mechanism Knowledge

The above review of work done to understand the mechanism of corrosion illustrates the need for continued study. It is generally accepted that there are three stages to pitting however there is much disagreement in the phenomenon behind each stage. Pits can be initiated in many different ways and the growth of pits can be attributed to different phenomenon. The reasons for pit repassivation are also not well understood. For these reasons, pit mechanism had been assigned a depth score of 3 and a breadth of 3.

Modeling of Pitting Corrosion Rates

The rate of corrosion of pits is an integral part of predicting pitting behavior and assessing remaining life of assets susceptible to pitting corrosion. The following summarizes the current understanding of pitting corrosion rates observed in the literature.

Modeling Pitting Rates in Piping under Insulation

Developing a model to predict CUI behavior in marine environments is needed to reduce failures, optimize maintenance and inspection schedules and aid in material selection for such applications. Pitting is a key degradation mechanism found in the field and a method for modeling the rate of pitting under insulation is needed.

Recommended Practice by Det Norske Veritas, DNV-RP-G101 (2002) uses degradation modeling to plan risk-based inspections. This method will be discussed in Section 2.6. The recommended practice includes a model for corrosion rate of carbon steel under insulation. It describes the rate as normally distributed and is a function of temperature. Table 3 below summarizes this rate.

Table 3: Corrosion rate (CR) determination for carbon steel under insulation from DNV-RP-G101 (2002)

Temperature (T)	Mean CR (mm/yr)	Standard deviation (mm/y)	Comment
<−5 °C			Probability of failure = 10^{-5}
−5 to 20 °C	Same as 20 °C	0.286	May overestimate rate, failures found at low temperatures
20–150 °C	$0.0067 \times T + 0.3$	0.286	
>150 °C			Refer to a specialist

This model assumes that if the insulation is wetted by salt water, these rates will apply and if insulation is not wet, there will be no CUI. This model is not specific to pitting and is a general corrosion rate model for CUI in carbon steel. This recommended practice does not include a corrosion rate model for stainless steel under insulation, the effects of CUI for stainless steel are accounted for using a probability of failure (PF) model that is included in Section 2.5.

Pitting is an issue under insulation however, no other information was found to indicate studies in this specific situation. As no information was found, the models for pitting corrosion in other situations summarized below may be used as a guide toward the development of a model for pitting rates of assets under insulation.

Predicting Pitting Rates

In operation, the depth of pitting is the most important characteristic that needs to be modeled. It is the depth of a pit that will effect containment and structural integrity of pipes and other components in marine environments.

A validated deterministic model for predicting pitting rates has not been found due to the complexity of the contributing factors and the apparently random nature of the process.

One study (Svintradze & Pidaparti, 2010) developed a governing equation for corrosion degradation due to pitting. This model was derived from solid state physics and attempted to model pit radius over time. This model included parameters that the authors were not able to determine and they recommended that further experimental work be conducted to validate their model.

Engelhardt, Urquidi-Macdonald, and Macdonald (1997) proposed a method that calculates damage functions for different types of localized corrosion types (pitting, crevice and stress corrosion cracking). This method is the only one found that allows for environmental conditions that change with time (corrosion potential, temperature, electrolyte composition, etc.). Using the damage function they suggest extrapolating short term experimental data to service life using extreme value statistics. They also argued that damage function analysis is an effective method for predicting future corrosion damage and indicate that updating the model with inspection data will improve the model.

The model depends on understanding four independent functions, the rate of defect nucleation, growth rate of the defect, rate of transition of one kind of defect to another, and the transition of an active pit into a passive pit or the transition of a pit into a crack.

To determine the rate of pit nucleation the point defect model can be used. This model includes external conditions of temperature, pH, metal potential, and halide ion activity to determine pit nucleation rate. To determine pit growth rate, they use an interpolation equation rather than the simple power law equation$L = A \cdot t^B$ (Engelhardt et al., 1997).

Valor, Caleyo, Alfonso, Rivas, and Hallen (2007) proposed a stochastic model to simulate pitting corrosion by combining pit initiation (Weibull function) and pit growth (non-homogeneous Markov process). They used extreme value statistics (Gumbel distribution) to determine maximum pit depth for extended periods of time. The authors validated their model using published data however; their method has been called into question by the original publisher of the data. Melchers (2007) argued that that the model is not appropriate for extrapolation from short term experimental data to long-term exposure because of its dependence on the power function for pit depth.

Stochastic models most commonly use extreme value distributions as maximum pit depth to be conservative and prevent leaking.

The Markov chain approach has been used to model pitting corrosion under the assumption that pitting damage is memory-less and current state alone determines future behavior (Caleyo, Velázquez, Valor, & Hallen, 2009).

Provan and Rodriguez (1989) developed a Markov stochastic process to model pit growth with time. The system was modeled by a discrete-space, continuous-parameter Markov process. They applied extreme value statistics to predict the deepest pit and found that if the maximum pit on an area is in one state $(j - 1)$ at time t, then during a time interval $(t + t)$, the pit grow to the next state (j) with probability

$$\lambda(j-1)[1+\lambda t/1+\lambda tk]\Delta t$$

where , k are corrosion system dependant and are based on short term experimental data.

Caleyo et al. (2009) used a continuous-time, non-homogenous linear growth Markov process to model external pitting corrosion in underground pipelines.

In other work, Melchers, 2003, Melchers, 2004, Melchers, 2008a and Melchers, 2008b and Melchers and Jeffrey (2008) has developed a model for corrosion in marine environments (immersion and atmospheric) that shows distinct phases, each with different corrosion rates based on the driving corrosion mechanism. Figure 6 illustrates this model.

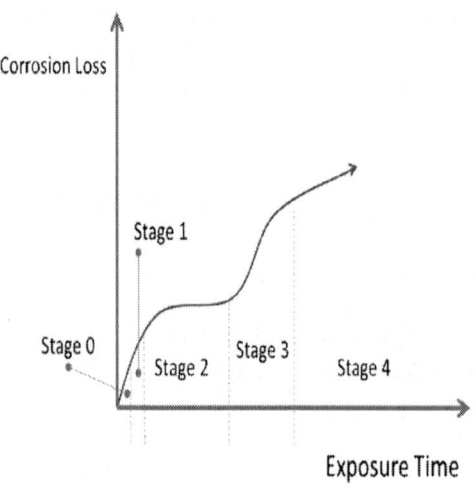

Figure 6: General schematic of model for corrosion loss showing the changing behavior of the corrosion process as a series of sequential phases adapted from Melchers (2003).

This model suggests that the conventional model for corrosion loss, $C(t) = A \cdot t^B$, is not applicable for the life of the component. The conventional model is based on diffusion of oxygen through increasingly thick corrosion layers and does not take into account changes in corrosion mechanisms with time (Roberge, 2008).

Melcher has demonstrated that this phenomenological model is applicable to pitting corrosion in marine immersion. His model shows five distinct stages, each with different pitting rate (pit depth/time) based on the driving corrosion mechanism (Melchers, 2004).

In Figure 6 above stage 0 is due to water velocity and surface finish, stage 1 is kinetic phase limited by oxygen diffusion through adjacent water, stage 2 is controlled by the rate of oxygen through corrosion product, stage 3 is rapid corrosion under anaerobic conditions, and stage 4 approximates steady state corrosion under anaerobic conditions. Parameters for this model were determined from long-term experimental field data.

Analysis of Pitting Rate Prediction Knowledge

As discussed above, there is little agreement on modeling of pitting corrosion behavior in marine applications. Pit depth had been identified by most as the key parameter to describe the rate of pitting and there have been many attempts to model this behavior. Due to the lack of consensus, correlation and validation, the score for pit modeling is low for both depth and breadth at 2.

Prediction of Asset Life (Fitness of Service) under Pitting Attack

Corrosion rates for a particular material in a specific environment can be used to make predictions about the life of an asset. The likelihood that an asset will continue to perform its function can be assessed in a variety of methods. Fitness-for-service (FFS) assessments are a common method to make these assessments. Evaluation of asset life can be made before a component is put into service to assess manufacturing or after to assess in-service damage (Holtam, Baxter, Ashcroft, & Thomson, 2011).

Many studies have predicted asset life or remaining life of components. In this work, we are primarily concerned with prediction of asset life under pitting corrosion attack. In the literature, pit density and maximum pit depth are the most important characteristics needed for assessment of component remaining life.

Standards and Recommended Practices

Holtam et al. (2011) surveyed FFS trends in industry to understand the application of FFS across industries. The survey found that API 579-1 (2007) standard, "Fitness-for-Service" was the most frequently used standard and that corrosion and erosion damage mechanisms were the most frequent procedures used within any standard.

API 579-1/ASME FFS-1 standard, Fitness-for-Service, outlines the method to assess the remaining life of components. For this work, part 6 of this recommended practice is discussed for its applicability to pitting.

Part 6, *Assessment of pitting corrosion*, gives a step by step method to qualify an asset for continued service based on known pitting damage. The assessment is used to determine the course of action for the component in terms of: rerate, repair, or replace.

There are three levels of assessment and each has conditions that govern their applicability. Generally, level 1 assessment is carried out on the simplest components and as complexity increases, more detailed assessments are required (levels 2 and 3). The assessments may also be completed sequentially if a lower level does not produce satisfactory results.

All assessment levels require equipment design data, maintenance and operation history, and material properties.

For level 1 assessment, pit damage is classified by pitting charts to determine the grade of pitting (1–8) and for level 2 a representative site is chosen for assessment with a minimum of 10 pit-couples included.

The level 1 assessment uses maximum pit depth to determine a remaining strength factor that is used to determine if the asset is fit for continued service. If the asset does not pass *level 1* assessment, the component can be directly repaired or replaced, or a level 2 or 3 assessment needs to be conducted.

Level 2 assessments determine if there is remaining strength in the component in both the circumferential and longitudinal stress directions. If the component does not pass a level 2 assessment, again the options are to repair, replaces or conduct a level 3 assessments.

Level 3 uses numerical methods to assess complex components and indicates if a component is fit for continued service, needs to be replaces, or repaired.

The remaining life of the component can also be estimated using this standard following a maximum allowable working pressure of the undamaged component (MAWP) approach. This assessment uses a pit propagation rate (PPR) to estimate future damage and to estimate MAWP as damage progresses with time. PPR is not specified in the standard but indicates that "… a Pit Propagation Rate should be determined based on the environmental and operating conditions (API 579-1, 2007, pp. 6–12)

There are other recommended practices to assess fitness-for-service and asset life available that are not summarized in this work. Some of these standards are summarized in Table 4. This table also includes an assessment of how these standards specifically address pitting corrosion.

Table 4: Select recommended practices for FFS assessments including specific methods for addressing pitting corrosion

Standard/Recommended practice	Specific pitting corrosion assessment procedure included	Method for determining pitting rate included
API 579-1/ASME FFS-1: Fitness-for-Service (API 579-1, 2007)	Yes	No
BS 7910: Guide to methods for assessing the acceptability of flaws in metallic structures (BS 7910, 2005)	No	No
FITNET: European fitness-for-service network (FITNET, 2006)	No	No
ASME B31.G: Manual for determining the remaining strength of corroded pipelines (ASME B31.G, 2012)	No	No

API 579-1/ASME FFS-1 is the only reviewed standard that specifically addresses pitting corrosion and no standard was found that includes a method for determining the rate of pitting.

Recommended Practice DNV-RP-G101 (2002) uses degradation modeling to plan inspections. This method is discussed in Section 2.6. To predict asset life this recommended practice uses a probability of failure (PF) per unit wall thickness as a function of temperature for local corrosion and stress corrosion cracking. Figure 7 illustrates the method included in DNV-RP-G101 to find the PF for local corrosion under insulation.

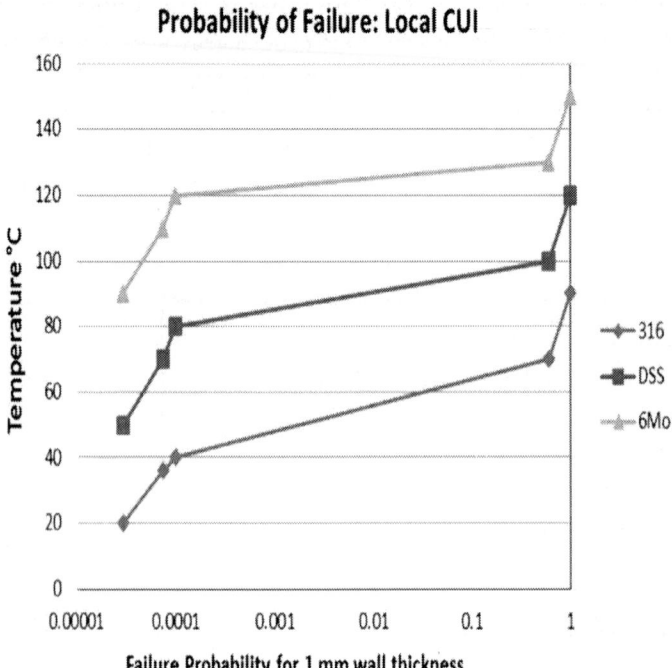

Figure 7: Schematic adapted from DNV-RP-G101 (2002) PF for local corrosion of stainless steel under insulation as a function of temperature.

In addition to the standards discussed above, there are many scholarly works that attempt to predict asset life of a component. As discussed in earlier sections, there is no consensus on modeling of pitting rate. For this reason, statistical models that predict the probability of failure of a component have been developed to minimize the effect of this issue.

Hodges et al. (2010) developed an internal system to assess corrosion risk that overcomes the lack of information available in practice. Their method incorporates data from many sources including engineering judgment. This method can then be used to plan monitoring systems and inspection schedules. The result of their work is a semi-quantitative risk assessment that they have shown to be useful to different assets and industries.

Others attempt to understand asset integrity modeling using uncertainty modeling (Akmar Mokh and Ismail, 2011, Race et al.,

2007 and Thodi et al., 2009). They found that pitting corrosion was most closely modeled using type1 extreme value and 3P-Weibull distributions. These models are then updated using Bayes theorem to assess risk to assets in service. They also incorporated inspection data into this model; this allows for the asset risk to be updated with new information and will lead to more realistic assessments of remaining asset life.

Other applications of asset integrity modeling for components susceptible to pitting may be useful to understanding CUI in marine applications.

Race et al. (2007) have developed a corrosion scoring model based on corrosion susceptibility and severity. They have developed this method considering three failure modes:

Probability of:

- Coating failure
- Cathodic protection failure
- Corrosion of unprotected pipe in soil environment

The model is developed by finding the probability of failure for each failure mode where

$$Probability\ of\ failure\ (PF) = Susceptibility\ factor \times Severity\ factor$$

- Coating failure PF = COATPF
- Cathodic protection failure PF = CPPF
- Corrosion of unprotected pipe in soil environment PF = SOILPF

These probability scores were determined through assessment of published data and engineering judgment. Combining all probability of failure scores led to a total failure score (TFS) for the system

$$TFS = \frac{COATPF + CPPF + SOILPF}{3}$$

The TFS is then fit to known data to determine a corrosion rate based on this score.

For this study (Race et al., 2007) the authors found

$$Maximum\ corrosion\ rate(mm/y) = 1.58 \times 10^{-4}\ TFS$$

In another study (Akmar Mokh & Ismail, 2011) the authors used the thinning failure function proposed byKhan, Haddara, & Bhattacharya

(2006) to assess failure of insulated piping. In this analysis, the variables are again assumed to be random, however, their distribution is assumed to be normal and the mean and standard deviation known. The failure probability is found using the following equation:

$$\mathbf{PF} = \int f_1(x_1) \ldots f_n(x_n) \mathrm{d}x_1 \ldots \mathrm{d}x_n$$

where $f_1(x_1)$ is the probability density function of each variable.

Using FORM to determine a reliability index   that satisfies the failure function and leads to a simplified function for failure probability for each defect. The probability of failure of the pipeline system can be found as follows:

PF(pipeline)$=1-\prod i_{(1-}$PF$_)$

The authors assume the defects are mutually exclusive. As this analysis has shown, the behavior of defects (pits) is complex and interaction and dependence of pits can be reasonably assured making this assumption by the authors questionable.

A case study was used to demonstrate usefulness of the function in assessing asset integrity using a corrosion rate that is assumed to be constant in time. The work by Melchers, 2003, Melchers, 2004,Melchers, 2007, Melchers, 2008a and Melchers, 2008b, and Melchers and Jeffrey (2008), has shown that corrosion behavior can vary significantly depending on the corrosion driving mechanism and that a constant corrosion rate is not always appropriate.

Analysis of Asset Life Prediction Knowledge Considering Pitting Corrosion

Asset integrity and fitness-for-service assessments are readily available and many additional procedures have been developed for specific industries and components. The analysis has shown that there is limited information on corrosion rates included in these methods. It is expected that corrosion rates are developed independently and then used in the analysis. As discussed in previous sections, there is currently no method for determining reliable long-term corrosion rates. For the purpose of this analysis, the depth of knowledge for this category is considered to be mid-level and is assigned a score of 6. This is due to the variety of different methods found indicating a lack of consensus and the lack of information on corrosion rates. The breadth of this category is considered high and given a score of 8. Fitness-for-service methods has been well demonstrated to be valid over many industries.

Risk-based Inspection

Risk-Based Inspection (RBI) is a methodology that develops inspection and maintenance plans based on risk. Risk is defined through analysis of the probability of an incident occurring and the severity of the consequences if an incident does occur. Using a risk-based inspection helps to focus inspection resources on key areas, evaluate the system wide risk against an operator set risk acceptance criteria, and develop optimal methods for inspection and monitoring (DNV-RP-G101, 2002).

Standard and Recommended Practices for RBI

The American Petroleum Institute (API) developed two recommended practices (RP) to address RBI:

- API Recommended Practice 580: Risk-Based Inspection (API RP 580, 2009)
- API Recommended Practice 581: Risk-Based Inspection Technology (API RP 581, 2008)

API 580 deals with defining RBI and instructing users on how to implement and sustain an RBI program. API 581 gives more specific procedures to develop an RBI program and to provide quantitative methods to assess overall plant risk. These two methods are intended to be used together and will be discussed together for this work.

API 580 defines terms and explains the basic concept of developing an RBI including overview of risk analysis, key elements of RBI programs, establishing boundaries, and data and information collection. It also introduces damage mechanisms and failure modes. These include corrosion, cracking, and metallurgical damage. Section 9.1.1 of API 580 lists general steps for identifying possible damage mechanisms. Once possible mechanisms are identified, Section 9.3 of API 580 describes how to assign an associated failure mode. These failure modes can include modes such as pinhole leaks, large leaks, or brittle fracture.

This information is used to complete a probability analysis. A general method for assessing probability of failure (PF) is discussed including qualitative and quantitative methods. Qualitative assessments are based on engineering judgment and then a description (high, medium, low or 0.1–0.01 times per year) is assigned. There are numerous quantitative approaches to PF indicated. Using probability shown as a distribution is one option; using manufacturer failure data is another.

To determine PF, API 580 lists the two main considerations; damage mechanism and rates, and the effectiveness of the inspection program. The steps to analyze these effects on PF are listed in the RP and summarized here (API RP 580, 2009):

- Identify active/credible damage
- Find damage susceptibility and rate
- Qualify inspection effectiveness
- Determine probability that the damage tolerance will be overcome.

Methods for determining these steps are generally discussed in API 580 and more in-depth in API RP 581 (2008). API 581 instructs a user on the calculations required to determine a PF for a component or system under study. API assesses PF as a combination of a generic failure frequency, a damage factor, and a management system factor.

The generic failure frequency is the basis of this assessment. It was set for different component types based on representative values from industry failure data. This failure rate is a baseline value before any damage occurs. The damage factor is applied to this baseline for each specific component and the management system factor is applied to all equipment.

Damage factors are determined by specific damage mechanisms. API 581 includes methods for determining damage factors for:

- Thinning (both general and local)
- Component Linings
- External Damage
- Internal Stress Corrosion Cracking
- High Temperature Hydrogen Attack
- Mechanical Fatigue (Piping Only)
- Brittle Fracture

The consequence of failure (CF) is then described and techniques for assessing introduced in API RP 580 (2009). These consequences are categorized as: safety and health impacts, environmental impacts, or economic impacts. Quantitative and qualitative techniques are introduced and can be measured in terms of safety or cost. API 581 includes methodologies for two levels of analysis. Level 1 analysis is a simplified method of evaluating the consequence of release of a limited number of fluids. This method includes determination of important system characteristics such as release rate, release hole size selection and flammable and explosive consequences.

Level 2 CF assessment provides a more detailed procedure for calculation. This analysis is used when the assumptions of the simplified level 1 assessment are not valid. An example of this situation given in the RP is when stored fluid is close to its critical point and the ideal gas assumption is invalid.

API 580 next generally discusses the risk assessment and management techniques. This section combines PF and CF to determine the risk. Risk = Probability (PF) × Consequence (CF).

API 580 gives information on prioritizing and evaluating acceptable risk and, using examples, demonstrates risk calculations and risk rankings. Once a risk tolerance is developed, the RP gives guidance on how to manage risks that are above the tolerance. Methods such as decommissioning, condition monitoring, and probability mitigation are discussed as ways to manage and reduce risk.

API 580 also gives information on reducing uncertainty in risk assessments through inspections. If damage mechanisms and rates of damage are assessed through inspection and then acted on, these methods can reduce PF and thus reduce overall risk.

API 581 again gives more detail and includes procedures for calculating risk. The RP includes equations for both area based risk and financial based risk.

The results of the risk assessment serve as a basis for developing the inspection plan. API 580 advises that the following be included in inspection procedure development:

• Risk criteria and ranking
• Risk drivers

- Asset history
- Number and results of inspections
- Type and effectiveness of inspections
- Equipment in similar service and remaining life

The type of inspection also plays a key role determining risk. Both API RP 580 (2009) and API RP 581 (2008) indicate that there are many factors that affect the risk. Some of these are; frequency of inspection, coverage of inspection, tools and techniques, procedures and practices, and inspection type (internal, on-stream, or external).

API 581 includes specific information on different types of components that may be included in RBI and gives specific advice on following the procedure for each. These components are: pressure vessels and piping, atmospheric storage tanks, pressure relief devices, and heat exchanger tube bundles.

API 580 and 581 do not contain specific sections on pitting corrosion. Pitting is mentioned throughout as a damage mechanism and API 581 indicates that damage rates are increased over general corrosion rates due to pitting for many situations.

- "These rates are 10 times the general corrosion rates to account for localized pitting corrosion" (API RP 581, 2008, pp. 2.B-13)
- "As a rule of thumb for carbon steel, the pitting rate is a factor of 5–10 times the coupon general corrosion rate, (calculated by weight loss)." (API RP 581, 2008, pp. 2.B-98)

DNV-RP-G101 (2002) "Risk Based Inspection of Offshore Topsides Static Mechanical Equipment" is another example of RBI recommended practice (RP). It describes the methodology for developing a RBI in an offshore production facility. This RP begins with a risk screening to categorize equipment into high, medium, or low risk. The second part of the process is a detailed quantitative assessment of higher risk areas.

The RP includes a guide to the screening process and recommends that this qualitative analysis be carried out by qualified knowledgeable personnel. This guide allows the assessment team to use engineering judgment to identify the consequence and probability (and thus the risk) as high or low for each component or system (DNV-RP-G101, 2002). Those components/systems that include a high rating for consequence and probability are further assessed in the detailed analysis. The RP indicates that if there is any question to the rating (high or low) the component/system should be included for further detailed study. Items that are found to have low or medium risk are followed up with maintenance activities and not considered further in the RBI methodology.

The detailed analysis calculates inspection schedules and techniques based on identified degradation mechanisms and current state of damage. This analysis aims to ensure that risk levels do not exceed a pre-determined acceptable risk limit.

This RP gives methods for determining the probability of failure (PF) and consequence of failure (CF) and illustrates the methods for combining to determine risk. Consequence modeling can be based on other analysis (quantitative risk analysis (QRA) or risk assessment methods (RAM)) but the RP also includes simplified methods to assess consequences. Event trees are recommended to identify possible consequence and to determine the CF. The practice separates consequence modeling into ignited and un-ignited consequences and outlines each in terms of personal safety, economic consequences, and environmental consequences.

Probability of failure modeling assesses the likelihood degradation mechanisms, the current PF and determines the PF as it changes with time. This PF will establish inspection intervals. The PF limit is determined from the acceptable risk limit and the CF.

Degradation mechanisms are identified as either an insignificant model, a susceptibility model, or a rate model. An insignificant model is used on specific material/fluid combinations and is considered to be fixed at $PF = 10^{-5}$/year. Inspection is considered irrelevant for this model (DNV-RP-G101, 2002). The susceptibility model determines PF based on operating conditions. This model is considered to be constant over time for given conditions. Inspection can be used to monitor process parameters. Rate models indicate that damage increases with time. Appendix C of DNV-RP-G101 includes typical material/fluid combinations and gives methods for determining PF for these situations. Two CUI models were discussed in Sections 2.4.2 and 2.5.1. Other models described include CO_2 model, microbial corrosion, corrosion based on water characteristics, and atmospheric corrosion. More detailed modeling of damage rates is suggested and probabilistic methods are suggested to obtain more accurate results.

Results from all analysis are combined and inspection is carried out to keep the risk of failure below the risk tolerance limit.

The above summarized recommended practices have been used as a guide or adapted by many researchers (Khalifa et al., 2012 and Khan et al., 2006). These works build on established procedure to improve these practices, increase safety and reduce cost.

Two state functions were developed by Khan et al. (2006) to describe material degradation. The first is for thinning of carbon steel and copper piping that measures the resistance of the material to applied stress.

$$g_t = S\left(1 - \frac{C \times \Delta t}{d}\right) - \left(\frac{P \times D}{2 \times d}\right)$$

where, S, Material Strength; C, corrosion rate; t, time increment; d, material thickness; P, operating pressure; D, diameter of the component.

A state function for stress corrosion cracking was also developed based on Paris's crack growth law (Khan et al., 2006).

$$g_c = K_{IC} - Y\left(\frac{P \times D}{2 \times d} + S\right)\sqrt{\pi\left(\frac{C_{cr}\Delta t^n}{R_{1/a}}\right)}$$

where, K_{IC}, material fracture toughness; Y, dimensionless geometric factor; S, residual stress; C_{cr}, crack growth rate; R/a, crack to length to-depth ratio.

The variables in the above models (S, C, P, K_{IC}, Y, R, A) have uncertainty and are considered to be random, making the material degradation process stochastic. These random variables are assumed to be independent and exponentially distributed leading to a gamma distributed process (Khan et al., 2006). Corrosion rate for the thinning model is assumed to be a linear function with time allowing the shape parameter ( ) of the gamma distribution to become $_0 t$. This leads to a failure distribution function for cumulative material degradation.

$$f_{x(t)}(x) = \frac{\beta^{\alpha_0 t}}{\Gamma(\alpha_0 t)}(x)^{\alpha_0 t - 1}e^{-\beta x} \quad \text{for } x > 0.$$

This function is then used for inspection updating using new inspection data and Bayesian updating (Khan et al., 2006).

Datla, Jyrkama, and Pandey (2008) introduced a probabilistic model of steam generator tube pitting corrosion based on inspection data from a nuclear generating station. A stochastic non-homogeneous Poisson process with pit size as a random variable was used. Their model was based on inspection data of pits that were greater than 50% of thickness.

Intensity function (Non-homogenous Poisson process):

λ(t)=αtβ−1

Expected number of pits (Poisson process):

$$E[N(t)] = \Lambda(t)\frac{\alpha}{\beta}t^{\beta} \quad \alpha = \text{scale parameter}, \beta = \text{shape parameter}$$

Pit depth distribution (Generalized Pareto Distribution):

$$F_x(x) = 1 - \left(1 - \xi\frac{x - \mu}{\sigma}\right)^{\frac{1}{\xi}} \quad \sigma = \text{scale parameter},$$

$\xi = $ shape parameter

Extreme pit depth distribution using extreme value theory:

FY(z)=exp(−Λz(t))

Inspection data was used to estimate parameters (α, β, λ, ξ, σ, μ).

In other works, Khalifa et al. (2012) describe a methodology used to develop a prediction tool to estimate the inspection sample size needed to determine the maximum localized corrosion depth of a process population. As inspection of all component area is not feasible, inspecting a limited number of sites may be necessary. To ensure the data collected from limited inspections represents the behavior of the entire system, the correct number of samples needs to be inspected. They developed a new method to determine the required sample size to assess localized corrosion. This method assumes that collected data is independent, has negligible measurement error, and follows the Gumbel extreme value distribution (Khalifa et al., 2012).

The method first divides the process under investigation into corrosion circuits. Each circuit includes components of the same material that are subjected to the same environment. The expectation is that areas in each corrosion circuit will experience the same degradation mechanisms.

The new equation to determine sample size needed to find the maximum localized corrosion was then demonstrated through a case study. Inspection data was used and a sample size determined using the methodology. This sample size was similar to that predicted by the proposed equation.

Analysis of RBI Using Pitting Corrosion Rate Modeling Knowledge

Risk-based inspection is well documented and is becoming more and more standard practice in industry. Methodologies are available to guide users through probability and consequence modeling. The recommended practices examined here indicate a well-established procedure for RBI. The RP reviewed include simplified methods for probability and consequence modeling. A need for precise modeling has been identified and the above papers indicate that this work is ongoing. There remains little information on probability modeling for pitting corrosion in general and less for pitting corrosion under insulation.

The depth of understanding of RBI is considered high at 7 as the RBI procedures are well-established and standards exist to guide in RBI development. It is the PF elements needed for RBIs that need to be further studied to increase accuracy. Pitting corrosion needs to be incorporated more directly and specifically. The breadth of understanding of this is considered mid-range at 6 as there are many standards used by different industries and there is limited consensus on these methods.

ANALYSIS AND DISCUSSION

Current State of Pitting Corrosion Knowledge

Earlier sections show that while we do understand a lot about pitting corrosion, there are still many aspects of this degradation that are relatively unknown or less known. Figure 8 below summarizes the assigned depth and breadth rating for the understanding of pitting corrosion in marine applications.

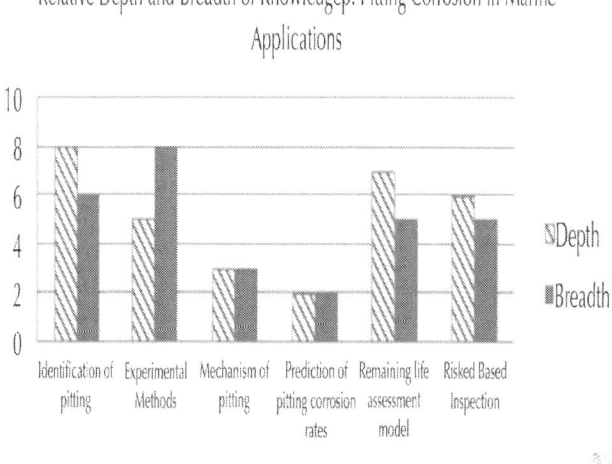

Figure 8: Relative depth and breadth of knowledge: Pitting corrosion in marine applications.

This graph clearly shows that the categories that need the most work to enhance our understanding of pitting corrosion are i) the mechanism of pitting and ii) the prediction of pitting corrosion. Figure 9 below again demonstrates the difference in knowledge of the separate categories of pitting knowledge.

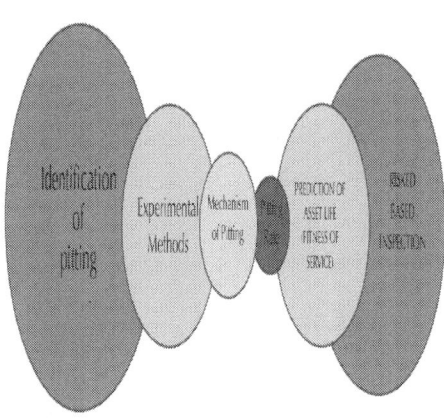

Figure 9: Relative depth of pitting corrosion knowledge.

The depth of knowledge of pitting corrosion modeling is significantly smaller than the other pitting categories analyzed. This lack of knowledge is limiting because as shown in Sections 2.5 and 2.6, pitting rate modeling is key to making accurate and reliable assessments. This is essential to improve safety and lower costs through FFS and RBI inspections.

Understanding the mechanism of pitting is also important as increased understanding of the phenomenon will make modeling more realistic. The methods described in Section 2.2 can be adapted along with the identification techniques described in Section 2.1 to allow for more research and data collection that will ultimately strengthen our understanding of all categories. These relationships are illustrated in Figure 10below.

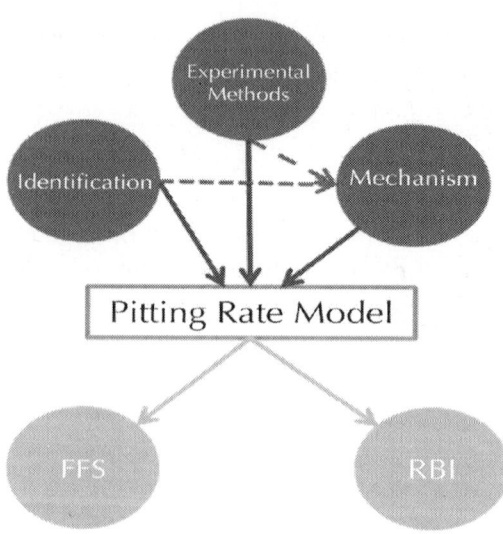

Figure 10: Relationship between pitting corrosion categories.

This figure illustrates the interdependence of the six analyzed categories. Pitting identification and experimental methods to study pitting are used to further understand and define pitting mechanism. All three of these categories are needed to develop a pitting rate model. This model will then be incorporated into FFS and RBI analysis to improve safety and reduce costs in marine operations.

Pitting under Insulation

Review of current models available for pipe failure due to corrosion and for corrosion rate has shown that there is no model available that to help model corrosion rates for piping systems under insulation in marine environments.

There are models available that may predict the failure rate of insulated pipes once a precise model of CUI corrosion rate is known.

There are models of corrosion rate available for other environments. These models indicate the importance of different variables that play a critical role in CUI modeling and thus will be used in failure model development.

Future Direction

The development of a corrosion rate model for CUI in offshore environments can only be developed when the variables that affect this type of corrosion are understood and the interactions between variables determined. The following fault tree shown in Figure 11 is used to help identify the base causes of CUI to begin modeling corrosion type for the purpose of fitness-for-service, RBI and failure modeling.

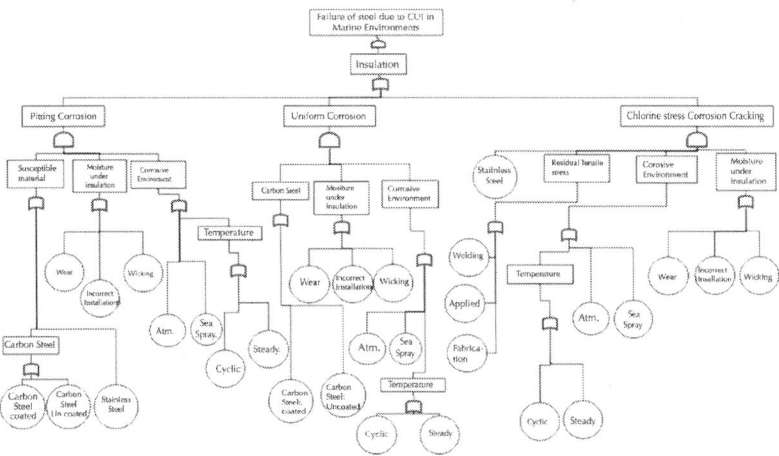

Figure 11: Fault tree of CUI of steel in marine environments (NOTE: Atm. = Atmosphere).

Some of the important causes of CUI in marine environments that need to be studied for their effect on corrosion rates have been identified. These include:

1.   Predicting moisture penetration under insulation including:
   - Wicking properties of insulation
   - Degradation of insulation over time
   - Installation issues (Human Factors)
2.   Effect of the moisture capturing:
   - Time of wetness (TOW)
   - Chlorine concentration [Cl]
   - Sulfur dioxide concentration [SO_2]
   - Rainfall amount
3.   Effects of cyclic temperatures
4.   Stress induced on stainless steel components.

Bacterial corrosion as a factor for long-term corrosion in marine atmospheres was introduced by Melchers (2004) and Melchers and Jeffrey (2008). Bacterial corrosion is sometimes called microbiological influenced corrosion (MIB). No published research is found that investigates this specific corrosion type under insulation. Further study is needed to determine if MIB is a factor along with pitting, uniform and stress corrosion cracking for CUI.

Further research and testing is needed to determine interactions of different parameters and to develop a corrosion rate model for CUI that can be used to predict reliability and fitness-for-service in marine environments.

Research Direction

To begin developing a model for CUI that includes pitting corrosion new field experiment set-up is to be developed. This is important to better understand mechanisms of pitting under insulation and to link this data to engineering design and analysis.

Long-term, periodic data collection is needed to determine corrosion rates under insulation. This data is not available from in-situ monitoring in industry due to issues with insulation removal, inspection techniques, and available resources. No field studies have been conducted to attempt to evaluate CUI in marine atmospheric conditions outside of a laboratory.

Most corrosion rates currently used in predictive models are based on short term laboratory corrosion data and no physical justification for the typical power function relationship is known (Engelhardt and Macdonald, 2004 and Melchers, 2004).

To generate long-term data, field testing is suggested. This would allow the collection of relevant environmental data and understanding of degradation mechanisms and corrosion rates. This would also create an opportunity to develop new inspection techniques to help determine pitting rates from on-line monitoring.

Field data could also be used to develop accelerated lab scale testing to further understand the long-term phenomenon and to simulate long-term exposure at specific environments and develop a pitting rate model.

CONCLUSION

This paper has summarized six categories of pitting corrosion and determined the state of understanding for each. It was found that the depth of knowledge of pitting corrosion rate modeling and pitting mechanism are significantly less than the other pitting categories analyzed.

From this work the following conclusions indicating the current state of knowledge can be made:

- There is significant information available on pit identification techniques and that these techniques are well understood.
- Well-established experimental methods are available to compare pitting resistance of materials.
- It is generally accepted that there are three stages to pitting.
- Pits can be initiated in many different ways and the growth of pits can be attributed to different phenomenon.
- Pit depth has been identified as the key parameter to describe pitting rate.

- Asset integrity and fitness-for-service (FFS) assessments are readily available and many additional procedures have been developed for specific industries and components.
- Risk-based inspection (RBI) is well documented and is becoming standard practice in industry.
- Reviewed RBI recommended practices include simplified methods for probability and consequence modeling.
- Through this analysis, the following conclusions indicating the need for further study can be made:
- Standard laboratory methods that accurately determine pitting rate are not available.
- Field data has shown that short term testing cannot be relied on to predict long-term corrosion.
- There is no consensus in the phenomenon of each of the three stages of pitting.
- The reasons for pit repassivation are also not well understood.
- There is little agreement on modeling of pitting corrosion rate.
- No FFS assessment found includes a method for determining pitting rate.
- Little information on probability modeling for pitting corrosion in general is available; less for pitting corrosion under insulation
- Categories that need the most work to complete understanding of pitting corrosion are the mechanism of pitting and the prediction of pitting corrosion.
- Pitting models are needed for more accurate FFS and RBI assessments.
- New experimental methods are needed to develop additional information on pitting.
- New inspection techniques could help to determine pitting rates from on-line monitoring.

REFERENCES

1.   Acevedo-Hurtado, P. O., Sundaram, P. A., Caceres-Valencia, P. G., Fachini, E. R., Miller, C. E., & Placzankis, B. E. (2008).

Characterization of atmospheric corrosion in Al/Ag lap joints. Corrosion Science, 50(11), 3123e3131. http://dx.doi.org. qe2a-proxy.mun.ca/10.1016/j.corsci.2008.08.014.

2. Akmar Mokh, A., & Ismail, M. C. (2011). Probabilistic reliability assessment of an insulated piping in the presence of corrosion defects. Journal of Applied Sciences (Asian Network for Scientific Information), 11(11), 2063e2067.

3. API 579-1/ASME FFS-1. (2007). Fitness-for-service. Washington D.C., USA: American

4. Petroleum Institute.

5. API RP 580. (2009). API recommended practice 580: Risk-based inspection. Washington, DC, USA: American Petroleum Institute.

6. API RP 581. (2008). API recommended practice 581: Risk-based inspection technology. Washington, DC, USA: American Petroleum Institute.

7. ASME B31.G. (2012). Manual for determining the remaining strength of corroded pipelines. Two Park Avenue, New York, NY, USA: American Society of Mechanical

8. Engineers.

9. ASTM B117-11. (2011). Standard practice for operating salt spray (fog) apparatus. West

10. Conshohocken, PA, USA: ASTM International.

11. ASTM B826. (2009). Standard test method for monitoring atmospheric corrosion tests by electrical resistance probes. West Conshohocken, PA, USA: ASTM International.

12. ASTM E3. (2011). Standard guide for preparation of metallographic specimens. West Conshohocken, PA, USA: ASTM International.

13. ASTM E1417. (2013). Standard practice for liquid penetrant testing. West Conshohocken, PA, USA: ASTM International.

14. ASTM F746. (2009). Standard test method for pitting or crevice corrosion of metallic surgical implant materials. West Conshohocken, PA, USA: ASTM International.

15. ASTM G1-03. (2011). Standard practice for preparing, cleaning, and evaluating corrosion test specimens. West Conshohocken, PA, USA: ASTM International.

16. ASTM G33. (2010). Standard practice for recording data from atmospheric corrosion tests of metallic coated steel specimens steels and related alloys by use of ferric chloride solution. West Conshohocken, PA, USA: ASTM International.

17. ASTM G46. (2005). Standard guide for examination and evaluation of pitting corrosion. West Conshohocken, PA, USA: ASTM International.

18. ASTM G48-11. (2011). Standard test methods for pitting and crevice corrosion resistance of stainless steels and related alloys by use of ferric chloride solution. West Conshohocken, PA, USA: ASTM International.

19. ASTM G50. (2010). Standard practice for conducting atmospheric corrosion tests on metals. West Conshohocken, PA, USA: ASTM International.

20. ASTM G61-86. (2009). Standard test method for conducting cyclic potentiodynamic polarization measurements for localized corrosion susceptibility of iron-, nickel-, or cobalt-based alloys. West Conshohocken, PA, USA: ASTM International.

21. ASTM G85. (2011). Standard practice for modified salt spray (fog) testing. West Conshohocken, PA, USA: ASTM International.

22. ASTM G150-99. (2010). Standard test method for electrochemical critical pitting temperature testing of stainless steels. West Conshohocken, PA, USA: ASTM International.

23. Baboian, R. (2005). Corrosion tests and standards: Application and interpretation. West Conshohocken, PA, USA: ASTM International.

24. Borucki, J. S. (1989). Liquid penetrant inspection, Nondestructive evaluation and quality control (Vol. 17). Materials Park, Ohio, USA: ASM International.

25. BS 7910. (2005). Guide to methods for assessing the acceptability of flaws in metallic structures. UK: British Standards Institution.

26. Byars, H. G. (1999). Inspection of surface equipment. Corrosion control in petroleum production. TPC publication 5 (2nd ed.). NACE International.

27. Caleyo, F., Velázquez, J. C., Valor, A., & Hallen, J. M. (2009). Markov chain modelling of pitting corrosion in underground

pipelines. Corrosion Science, 51(9), 2197e 2207. http://dx.doi.org.qe2a-proxy.mun.ca/10.1016/j.corsci.2009.06.014.

28. Chaves, I. A., & Melchers, R. E. (2012). External corrosion of carbon steel pipeline weld zones. In Paper presented at the Rhodes, Greece.

29. Clarke, A., & Eberhardt, C. (2002). Principles of confocal laser scanning microscopy. Microscopy techniques for materials science. Sawston, Cambridge, UK: Principles of Confocal Laser Scanning Microscopy.

30. Courbot, A., Nasr, A., Gilmour, W., & Biedermann, C. (2013). A new approach to pipeline inspection using autonomous underwater vehicles. In Paper presented at the 2013 offshore technology conference, http://dx.doi.org/10.4043/24224-MS.

31. Craig, B. A. (1991). Films and pitting corrosion. Fundamental aspects of corrosion films in corrosion science. NY, USA: Plenum Press.

32. Cramer, S. D., & Jones, B. P. (2005). Planning and design of tests. In R. Baboian (Ed.), Corrosion tests and standards: Application and interpretation (2nd ed.). (pp. 49e 58) West Conshohocken, PA, USA: ASTM International.

33. Datla, S. V., Jyrkama, M. I., & Pandey, M. D. (2008). Probabilistic modelling of steam generator tube pitting corrosion. Nuclear Engineering and Design, 238(7), 1771e 1778. http://dx.doi.org/10.1016/j.nucengdes.2008.01.013.

34. Davies, M., & Scott, P. J. B. (2003). 8.1.1.4 inspection. Guide to the use of materials in waters. Houston, TX, USA: NACE International.

35. De-Abreu, Y., Clark, J. C., Helander, J., Manko, D., & Suarez, H. A. (2012). Pitting resistance evaluation of stainless steel alloys in presence of neat production chemicals by using accelerated electrochemical tests. In Paper presented at the NACE International, Corrosion 2012. Salt Lake City, UT, USA.

36. DNV-RP-G101. (2002). Recommended practice, risk based inspection of offshore topsides static mechanical equipment. Høvik, Oslo Country, Norway: Det Norske Veritas.

37. Eden, D. A., & Kane, R. D. (2005). Single electrode probes for on-line electrochemical monitoring. In Paper presented at the Corrosion 2005, NACE International. Houston, TX, USA.

38. Engelhardt, G., & Macdonald, D. D. (2004). Unification of the deterministic and statistical approaches for predicting localized corrosion damage. I. Theoretical foundation. Corrosion Science, 46(11), 2755e2780. http://dx.doi.org.qe2a-proxy. mun. ca/10.1016/j.corsci.2004.03.014.

39. Engelhardt, G., Urquidi-Macdonald, M., & Macdonald, D. D. (1997). A simplified method for estimating corrosion cavity growth rates. Corrosion Science, 39(3), 419e441. http://dx.doi. org/10.1016/S0010-938X(97)86095-7.

40. Estupiñán-López, F. H., Almeraya-Calderón, F., Bautista Margulis, R. G., Baltazar Zamora, M. A., Martínez-Villafañe, A., Uruchurtu, Ch. J., et al. (2011). Transient analysis of electrochemical noise for 316 and duplex 2205 stainless steels under pitting corrosion. International Journal of Electrochemical Science, 6, 1785e1796.

41. FITNET European fitness for service network(2006). Final technical report. Retrieved 05/13, 2013, from http://www.eurofitnet.org/ FITNETFinalTechnRepwithProj Management30Jan07.pdf.

42. Frankel, G. (1998). Pitting corrosion of metals: a review of the critical factors. Journal of the Electrochemical Society, 145(6), 2186e2198.

43. Frankel, G. (2008). Electrochemical techniques in corrosion: status, limitations, and needs. Journal of ASTM International, 5(2).

44. Gale, W. F., & Totemeier, T. C. (2004). Chapter 10: Metallography. Smithells metals reference book (8). Materials Park, OH, USA: ASM International.

45. Grubb, J. F., DeBold, T., & Fritz, J. D. (2005). Corrosion of wrought stainless steels. In Corrosion: Materials (Vol. 13B). Materials Park, Ohio, USA: ASM International.

46. Heerings, J., Trimborn, N., & den Herder, A. (2006). Inspection effectiveness and its effect on the integrity of pipework. In Asia-Pacific conference on NDT. Auckland, New Zealand.

47. Heidersbach, R. (2011). Inspection, monitoring, and testing. Metallurgy and corrosion control in oil and gas production. Hoboken, NJ, USA: John Wiley & Sons.

48. Hodges, S., Spicer, K., Barson, R., John, G., Oliver, K., & Tipton, E. (2010). High level corrosion risk assessment methodology for oil & gas systems. In Paper presented at the NACE International, Corrosion 2010. San Antonio, TX, USA.

49. Holme, B., & Lunder, O. (2007). Characterisation of pitting corrosion by white light interferometry. Corrosion Science, 49(2), 391e401. http://dx.doi.org.qe2a-proxy. mun.ca/10.1016/j. corsci.2006.04.022.

50. Holtam, C. M., Baxter, D. P., Ashcroft, I. A., & Thomson, R. C. (2011). A survey of fitness-for-service trends in industry. Journal of Pressure Vessel Technology, 133(1), 14001.

51. Jana, S. (1995). Non-destructive in-situ replication metallography. Journal of Materials Processing Technology, 49(1e2), 85e114.

52. Jasiczek, M., Kaczorowski, J., Kosieniak, E., & Innocenti, M. (2012). A new approach to characterization of gas turbine components affected by pitting corrosion. Journal of Failure Analysis and Prevention, 12(3), 305e313.

53. Jia, Z., Du, C., Li, C., Yi, Z., & Li, X. (2011). Study on pitting process of 316L stainless steel by means of staircase potential electrochemical impedance spectroscopy. International Journal of Minerals and Materials, 18(1), 48e52. http://dx.doi.org/ 10.1007/s12613-011-0398-9.

54. Jirarungsatian, C., & Prateepasen, A. (2010). Pitting and uniform corrosion source recognition using acoustic emission parameters. Corrosion Science, 52(1), 187e 197. http://dx.doi.org.qe2a-proxy.mun.ca/10.1016/j.corsci.2009.09.001.

55. JIS G 0578. (2000). JIS G 0578 Method of ferric chloride tests for stainless steels. Tokyo, Japan: Japanese Industrial Standard.

56. Jones, D. A. (1996). Pitting and crevis corrosion. Principles and prevention of corrosion. Saddle River, NJ, USA: Prentice Hall.

57. Khalifa, M., Khan, F., & Haddara, M. (2012). A methodology for calculating sample size to assess localized corrosion of process components. Journal of Loss Prevention in the Process Industries, 25(1), 70e80.

58. Khan, F. I., Haddara, M. M., & Bhattacharya, S. K. (2006). Risk-based integrity and inspection modeling (RBIIM) of process components/system. Risk Analysis, 26(1), 203e221.

59. Klassen, R. D., & Roberge, P. R. (2003). Role of solution resistance in measurements from atmospheric corrosivity sensors. In Paper presented at the Corrosion 2003. San Diego, CA, USA.

60. Krakowiak, S., & Darowicki, K. (2002). The effect of the temperature change rate on determination of the critical pitting temperature of stainless steels. AntiCorrosion Methods and Materials, 49(2), 105e110.

61. Kros, H. (2011). Performing detailed level 1 pipeline inspection in deep water with a remotely operated vehicle (ROV). In Paper presented at the offshore technology conference, http://dx.doi.org/10.4043/21969-MS.

62. Lothongkum, G., Vongbandit, P., & Nongluck, P. (2006). Experimental determination of E-pH diagrams for 316L stainless steel in air-saturated aqueous solutions containing 0e5,000 ppm of chloride using a potentiodynamic method. AntiCorrosion

63. Methods and Materials, 53(3), 169e174. http://dx.doi.org/10.1108/ 00035590610665581.

64. McIntyre, P., & Vogelsang, J. (2009). Progress in corrosion e The first 50 years of the EFC: (EFC 52). Leeds, UK: Maney Publishing.

65. Mcleod, D., Jacobson, J. R., & Tangirala, S. (2012). Autonomous inspection of subsea facilities-Gulf of Mexico trials. In Paper presented at the offshore technology conference. Houston, TX, USA, http://dx.doi.org/10.4043/23512-MS.

66. Melchers, R. E. (2003). Modeling of marine immersion corrosion for mild and lowalloy steels. Part 1. Phenomenological model. Corrosion, 59(4), 319.

67. Melchers, R. E. (2004). Pitting corrosion of mild steel in marine immersion environment. Part 1. Maximum pit depth. Corrosion, 60(09).

68. Melchers, R. E. (2007). Discussion on "Stochastic modeling of pitting corrosion: a new model for initiation and growth of multiple pits" by A. Valor, F. Caleyo, L. Alfonso, D. Rivas, J.M. Hallen [Corros. Sci. 49 (2007) 559]. Corrosion Science, 50(5), 1518e1519. http://dx.doi.org.qe2a-proxy.mun.ca/10.1016/j.corsci.2008.01. 017.

69. Melchers, R. (2008a). Modeling of long-term corrosion loss and pitting for chromium bearing and stainless steels in seawater. Corrosion, 64(2), 143e154.

70. Melchers, R. E. (2008b). A new interpretation of the corrosion loss processes for weathering steels in marine atmospheres. Corrosion Science, 50(12), 3446e 3454. http://dx.doi.org.qe2a-proxy.mun.ca/10.1016/j.corsci.2008.09.003.

71. Melchers, R. E., & Jeffrey, R. (2008). Modeling of long-term corrosion loss and pitting for chromium-bearing and stainless steels in seawater. Corrosion, 64(2), 143e 154. http://dx.doi.org/10.5006/1.3280683.

72. Moran, E., Frankel, G. S., & Kim, Y. H. (2011). Localized corrosion resistance of 21% Cr ferritic stainless steel. Corrosion, 67(9), D1.

73. Novak, P. (2007). Environmental deterioration of materials. Slovakia: WIT Press.

74. Papavinasam, S., Attard, M., Doiron, A., Demoz, A., & Rahimi, P. (2012). Non-intrusive techniques to monitor internal corrosion of oil and gas pipelines. In Paper presented at the NACE International, Corrosion 2012. Salt Lake City, UT, USA.

75. Pechacek, R. W. (2003). Advanced NDE methods of inspecting insulated vessels and piping for ID corrosion and corrosion under insulation (CUI). In Paper presented at the NACE International Corrosion 2003. San Diego, CA, USA.

76. Pellegrino, B. A., & Nugent, M. (2012). Nondestructive testing technologies and applications for detecting, sizing and monitoring corrosion/erosion damage in oil & gas assets. In Paper presented at the NACE International Corrosion 2012. Salt Lake City, UT, USA.

77. Phull, B. (2003). Evaluating pitting corrosion. Corrosion: Fundamentals, testing, and protection (Vol. 13A (pp. 545e548). Materials Park, Ohio, USA: ASM International.

78. Power, R. J., & Shirokoff, J. (2012). Techniques for in situ corrosion studies of 316L stainless steel in sulfuric acid solutions. Recent Patents on Corrosion Science, 2, 2e21.

79. Power, R. J., & Shirokoff, J. (2013). Techniques for in situ corrosion studies of 316L stainless steel in hydrometallurgical process

solutions. Recent Patents on Corrosion Science, 3(1), 1e11. http://dx.doi.org/10.2174/2210683911202999000.

80.   Provan, J. W., & Rodriguez, E. S., III (1989). Part I: Development of a Markov description of pitting corrosion. Corrosion, 45(3), 178e192.

81.   Race, J. M., Dawson, S. J., Stanley, L. M., & Kariyawasam, S. (2007). Development of a predictive model for pipeline external corrosion rates. Journal of Pipeline Engineering, 6(1), 13e30.

82.   Raman, A. (2007). Materials maintenancedNon-destructive testing of ME components, materials selection and applications in mechanical engineering. South Norwalk, CT, USA: Industrial Press Inc.

83.   Rao, B. P. C., Jayakumar, T., & Raj, B. (2007). Electromagnetic NDE techniques for materials characterization. In C. H. Chen (Ed.), Ultrasonic and advanced methods for nondestructive testing and material characterization. Hackensack, NJ, USA: World Scientific Publishing Co.

84.   Reardon, A. (2011). Metallography. Metallurgy for the non-metallurgist. West Conshohocken, PA, USA: ASM International.

85.   Reiner, L., & Bavarian, B. (2007). Thin film sensors in corrosion applications. In Paper presented at the NACE International, Corrosion 2007. Nashville, TN, USA.

86.   Roberge, P. R. (2007). Corrosion inspection and monitoring. Hoboken, New Jersey, USA: John Wiley & Sons.

87.   Roberge, P. R. (2008). Corrosion engineering: Principles and practices. New York, NY, USA: McGraw-Hill.

88.   Roberge, P. R. (2011). Corrosion monitoring. In R. Revie (Ed.), Uhlig's corrosion handbook (3rd ed.). (pp. 1181e1201) Hoboken, NJ, USA: John Wiley & Sons.

89.   Schumacher, M. (1979). Seawater corrosion handbook. Park Ridge, NJ, USA: William Andrew Publishing/Noyes.

90.   Shreir, L., Jarman, R., & Burstein, G. (1994). Corrosion testing, monitoring and inspection. In (3rd ed.).Corrosion (Vols. 1 and 2; pp. 19:154e19:178) Elsevier.

91.   Siow, K. S., Song, T. Y., & Qiu, J. H. (2001). Pitting corrosion of duplex stainless steels. Anti-Corrosion Methods and Materials, 48(1), 31e37.

92. Snow, G., & Shirokoff, J. (2008). Electrochemical computer controlled methods for measuring the corrosion resistance of stainless steel in industrial applications. In Presented at Newfoundland electrical and computer engineering conference (NECEC). St. John's NL, CAN.

93. Sorg, M., & Ladwein, T. (2009). Investigation of the pitting corrosion behaviour of stainless steels in ethanol containing fuels. In Paper presented at the NACE International, Corrosion 2009. Atlanta, Georgia, USA.

94. Suresh Kumar, M., Sujata, M., Venkataswamy, M. A., & Bhaumik, S. K. (2008). Failure analysis of a stainless steel pipeline. Engineering Failure Analysis, 15(5), 497e 504. http://dx.doi.org/10.1016/j.engfailanal.2007.05.002.

95. Svintradze, D. V., & Pidaparti, R. M. (2010). A theoretical model for metal corrosion degradation. International Journal of Corrosion, 2010.

96. Szklarska-Smialowska, Z. (2005). Pitting and crevice corrosion. Houston TX, USA: NACE International.

97. Takahashi, T., Shibano, J., Ishitsuka, K., & Kobayashi, M. (2012). Mechanism of pitting corrosion of welded AISI 304 under a freeze-thaw corrosive environment. In Paper presented at the NACE International, Corrosion 2012. Salt Lake City, UT, USA.

98. Terribile, A., Schiavon, R., Rossi, G., Zampato, M., & Indrigo, D. (2007). A remotely operated tanker inspection system (ROTIS). In Paper presented at the offshore Mediterranean conference and exhibition. Ravenna, Italy.

99. Thodi, P., Khan, F., & Haddara, M. (2009). The selection of corrosion prior distributions for risk based integrity modeling. Stochastic Environmental Research and Risk Assessment, 23(6), 783e809.

100. Valor, A., Caleyo, F., Alfonso, L., Rivas, D., & Hallen, J. M. (2007). Stochastic modeling of pitting corrosion: a new model for initiation and growth of multiple corrosion pits. Corrosion Science, 49(2), 559e579. http://dx.doi.org.qe2a-proxy.mun.ca/10.1016/j.corsci.2006.05.049.

101. Vander Voort, G. F. (2004). Metallography and microstructures. (1989). In ASM International (Ed.), ASM handbook (Vol. 9). Materials Park, OH, USA: ASM International.

102. Vander Voort, G. (1999). Metallography, principles and practice. Materials Park, OH, USA: ASM International.

103. Visual inspection, nondestructive evaluation and quality control. In ASM International (Ed.), ASM handbook (Vol. 17; pp. 3e11). ASM International.

104. Yu, S., Ura, T. (2002). Visual inspection of underwater structures by autonomous underwater vehicles based on positioning using artificial objects placed on them. In Paper presented at the fifth ISOPE Pacific/Asia offshore mechanics symposium. Daejeon, Korea.

Chapter 8

Improved FTA Methodology and Application to Subsea Pipeline Reliability Design

Jing Lin, Yongbo Yuan, and Mingyuan Zhang

Department of Construction Management, Dalian University of Technology, Dalian, China

ABSTRACT

An innovative logic tree, Failure Expansion Tree (FET), is proposed in this paper, which improves on traditional Fault Tree Analysis (FTA). It describes a different thinking approach for risk factor identification and reliability risk assessment. By providing a more comprehensive and objective methodology, the rather subjective nature of FTA node discovery is significantly reduced and the resulting mathematical calculations for quantitative analysis are greatly simplified. Applied to the Useful Life phase of a subsea pipeline engineering project, the approach provides a more structured analysis by constructing a tree following the laws of physics and geometry. Resulting improvements are summarized in comparison table form.

INTRODUCTION

Risk assessment is an important aspect of project management in civil engineering and other industrial fields. As its recognition has increased, various methodologies and more specific classifications were developed over the years to assist in risk identification. Reliability risk analysis is considered a part of risk assessment theory.

Reliability risk assessment should be applied throughout the entire lifecycle of a product or structure. Reliability is always the opposite of failure. Risk assessment attempts to quantify probability of failure and in addition the consequences of failure. Therefore analyzing failure modes and mechanisms has become an essential procedure, especially at the beginning of design when corrective actions are most easily incorporated. Failure analysis originated for reactive problem solving or trouble shooting in the manufacturing industry. Internationally, commonly used techniques are Event Tree Analysis (ETA), Fault Tree Analysis (FTA) [1], Failure Modes & Effects Analysis (FMEA), Checklists, What–If Analysis, Preliminary Hazard Analysis (PHA) [2], Cause-Consequences Analysis, Safety Review, Relative Ranking, Human Reliability, Hazard & Operability Analysis (HAZOP), etc. Basically their purpose is to discover and prevent product/structure malfunctions, ensure reliability during the lifecycle and prevent safety hazards while in service.

These approaches follow a standard procedure, starting with the failure description, and then generating hypotheses based on historical data or experience of experts. Hypotheses are grouped into different categories for further calculations and analysis, and finally guide implementation based on qualitative or quantitative conclusions.

In this paper, a new method of reliability risk assessment is developed and applied to a subsea pipeline system. Subsea pipelines are major oil and gas transportation facilities for the deep ocean hydrocarbon mining industry. Pipelines should be designed strong and reliable enough to survive complicated environmental undersea stresses as well as internal and external impacts from both nature and human activities. Impacts could be constant ocean waves, current flow, earthquake or other vibration, corrosion, etc. The consequences of pipeline leakage are destructive and catastrophic for marine life, followed by huge economic loss and environmental cost. For example, in the well-known British Petroleum accident in the Gulf of Mexico in 2010, about 240,000 barrels of crude oil spilled out of three leakage spots along the pipeline per day. 400 different species of life in the area were put at risk. The direct economic loss was over 1 billion USD. In the following year, 2011, an oil leakage disaster caused by a pipeline failure from an offshore oil well shared by ConocoPhillips and China National Offshore Oil Corporation occurred, which contaminated over 170 km^2 of ocean. It will take decades to clean up the resulting environmental pollution. Clearly it is extremely important to identify reliability failure modes and design them out, if possible, or at least minimize their impact, should they occur. Risk assessment for a subsea pipeline system includes risk factor identification and evaluation, risk control strategies, corrective actions, suitable parameter tolerances and risk monitoring. Such analyses help to build a reliable pipeline system in an effective and economic way.

Domestically in China, a few major methodologies have been applied to risk management of subsea pipeline systems. For example, FTA was developed for subsea pipeline system failure modes by YJ Xie [3]. Analytic Hierarchy Process (AHP) was fitted inside this analysis to evaluate an Expert Scoring Method for reducing data subjectivity. Fuzzy logic analysis has often been used to assess the risk level under various fuzzy conditions [4] and to do criticality assessment of the consequences after a subsea pipeline system fails [5]. [6] used the Fuzzy Relative Matrix method to configure a single factor fuzzy matrix to evaluate the overall safety level of a subsea pipeline system. Based on a commonly used risk assessment system proposed by W. Kent Muhlbauer [7], the root causes of historical subsea pipeline failures fall into four different categories: Third party destruction, Corrosion, Poor design and Operational mistakes. Within each category, further classifications are made for underlying factors. Each factor is rated within its given scope. The sum of the scores is calculated for evaluating the overall risk level of the subsea pipeline system.

A common inadequacy of these methodologies is the lack of sufficient objectivity and their potential for leakage (i.e., missed failure modes). A new method, Failure Expansion Tree (FET), is proposed here for design reference. Significant advantages are shown for its application to subsea pipelines.

MATERIALS AND METHODS

FTA and Traditional Application to Pipeline Reliability Design

The basic FTA method was originated by Bell Laboratories in 1962. It was developed for evaluating the security systems for rocket launching. After that, the airplane maker, Boeing, brought the method to a higher level, both qualitatively and quantitatively [8]. In the following decades, FTA has been recognized and used for reliability analysis and risk evaluation in many industries. For example, FTA has been widely applied in the aviation industry by the U.S. Federal Aviation Administration (FAA) since 1970 [9]. Following the nuclear incident at Three Mile Island, the U.S. Nuclear Regulatory Commission expanded probabilistic risk assessment research, including FTA [10]. In 1992, the United States Department of Labor Occupational Safety and Health Administration (OSHA) published its Process Safety Management (PSM) standard. In 19 CFR 1910.119, FTA was officially accepted as a method for process hazard analysis [11], etc.

FTA is regarded as an efficient way to describe cause-effect relationships using a logic diagram. It is the starting point for qualitative and quantitative analysis of failure modes. There are five steps to build a FTA: (1) define the problem and classify its boundary, (2) construct the tree, (3) collect the minimum cut/route set, (4) perform qualitative analysis and (5) perform quantitative analysis. The system failure is on the top level of the tree followed by the direct causes (sub-event/sub-factors) on the second level, A1, A2, etc. The two levels are connected by appropriate "And" and "Or" logic gates. The same scheme is used to work further down the tree along each branch until reaching the point that the event/feature cannot be divided anymore. The elements on the very bottom of the tree are called basic elements. Fig. 1 shows a flow chart of the development sequence of a FTA.

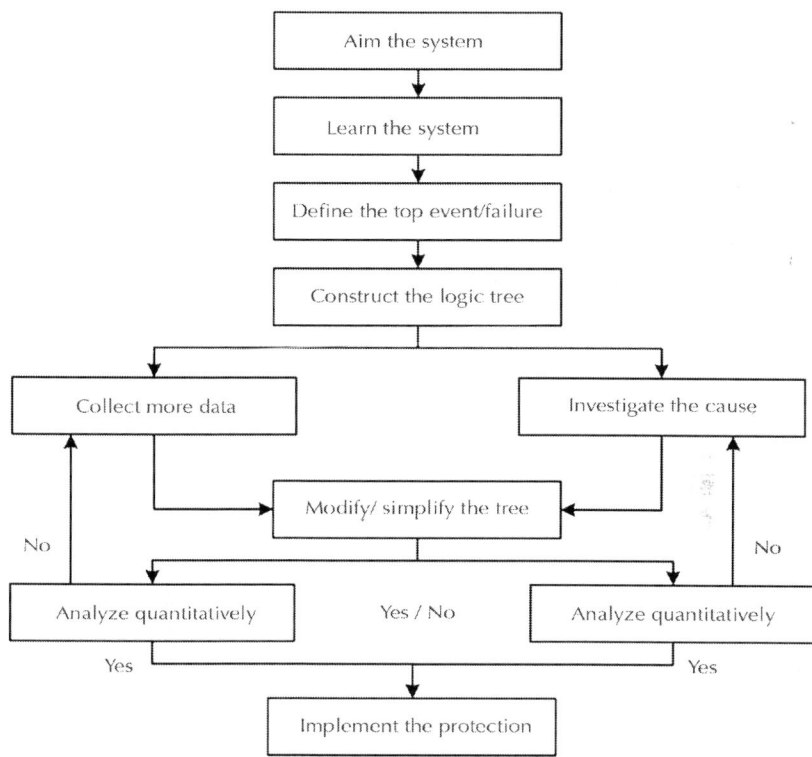

Figure 1: Process Flow Diagram for development of a FTA.

Besides [3], FTA examples have been mentioned and demonstrated by various research organizations and utilized in the construction of subsea pipelines. For example, fuzzy fault tree analysis was used for estimating the failure probability of oil and gas transmission pipelines [12]and for evaluating faults from third party damage [13]. In [14] it has been applied to risk assessment for a subsea pipeline under haphazard loads.

The Fault Tree Analysis example in [3] is introduced here to show how the FTA methodology has been applied to subsea pipeline systems. The tree, shown in Fig. 2, codes each failure factor with a number, F(n). A table of these codes is given in Table S2. Note how failure factors are combined with 'And' and 'Or' gates. Careful inspection of the FTA structure and its codes reveals certain weaknesses which could potentially result in failure mode leakage.

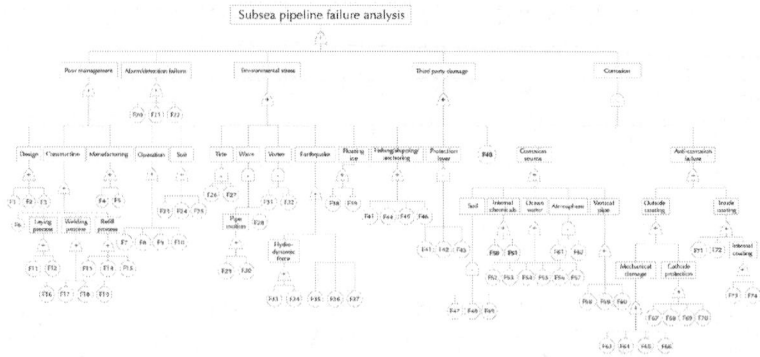

Figure 2: A typical Fault Tree analysis.This example for a Subsea Pipeline System is taken from [3].

Boxes/factors listed horizontally at a given level occasionally overlap each other, e.g., F23 (low soil viscosity) and F28 (high soil liquidity) are basically stating the same thing, so probability assignments for these "factors" are not independent and cannot be combined with "And" and "Or" gates. Cause and effect are confounded, e.g., F62 the result, "Corrosion in the atmosphere", and various initiating causes, F63–F66 (anti-corrosion layer damage during construction, shipping, installation and operation) are stated as separate factors. This causes a layer discrimination problem. Another example is the bottom node F67, "Improper cathode protection design". This statement creates much confusion. Cathode protection damage is NOT simply caused by design, but could be manufacturing, construction, or induced by various environmental stresses. Further inspection of the table reveals other types of logic problems which often result from brainstorming and subjective grouping of possible failure causes.

This example for a Subsea Pipeline System is taken from [3].

New Logic Tree Proposal

Tree Purposes and Objectives

The purpose of traditional FTA is to build a failure analysis model which is useful for both proactive risk control during the design stage and for reactive problem solving/trouble shooting during operating life. A good tree structure should lead to standard conclusions with judgments that are independent of tree builders. In this way, sound design considerations can be generated in a realistic and rigorous manner with minimal leakage during implementation.

Tree Design Principles

In order to develop a more complete and less subjective tree, a new methodology, Failure Expansion Tree (FET) analysis is proposed in which certain principles are strictly followed to minimize sensitivity of the bottom line results.

Principle #1: Failure elements (bottom line boxes/nodes) should be events or things physically observable on the structure. The term Failure Mode is defined as "the effect by which a failure is observed on a failed item" [15]. This most obvious and simplest rule has been frequently misused or ignored in daily engineering work. As a result, a misnamed failure could lead the decision maker to incorrect corrective actions and improper sublevel splits. For example, in Fig. 2, in the first level, "poor management" is not a Failure Mode. It is a subjective conclusion drawn by engineer/expert when a failure occurs. The correct description ought to be something directly observed on a physical part. In fact, in this case, "poor management" should not have been listed as a sub-branch. Similar misnaming problems (not all listed) need improvement in this FTA based on Principle #1. By keeping a "common language" between design and problem solving engineers, there would be less confusion and fewer bad assumptions based on differences in understanding.

Principle #2: No bias/prejudgment. Due to prior expert knowledge it is often difficult to draw conclusions without any bias or prejudgment. During failure analysis, instead of beginning with the appearance of the failure directly from the structure and working backwards to its cause, it is easy to leap directly to causes by skipping over the intervening chains-of-events which led to the failure. This ends up mixing failure modes and effects, which then creates difficulty and confusion for further splits in the tree. Taking the example mentioned in Section 2, F65 "Anti-corrosion layer damage", is the result of "mechanical damage", instead of the cause of it. It should have been removed from this level. Other "Mechanical damage" nodes (i.e., F63, "Anti-corrosion layer defect caused during construction process, F64, "Anti-corrosion layer defect caused during shipping", and F66, "Anti-corrosion layer damage caused during operation") should remain because they are all factors which lead to the mechanical damage. The categories of causes are based on the different physical aspects of the problem (we call these "basis of split"). Again, such modifications can be applied in many other places in Fig. 2 based on this principle.

Principle #3: Boxes must be mutually exclusive. In a logic split, the horizontal elements at the same level should be mutually exclusive to each other, in other words, independent of each other. There should be no overlap of functions or failures between elements within the same level. If such a split cannot be established, it means a higher level split was not properly made. Without this structural requirement interactions between basic elements could be missed later and quantitative calculations based on the tree cannot be correct.

Besides the example mentioned in Section 2 (F23, "low soil viscosity" and F28, "high soil liquidity"), all three splits under the branch of "Design" ("F1, "Poor design model", F2 "Improper safety parameter values, and F3 "Soil parameters don't match reality") have the same problem. They are confounded with each other, because design model includes safety parameters, and soil parameters. The boxes are not clearly separated. To make the improvement based on Principle #3, "Design" can be split into different aspects of it, such as product development stage, manufacturing stage, or operation stage, which do NOT have overlap functions of each other. And in each stage, different parameters are required to enhance the total reliability of the structure.

Principle #4: Boxes must be collectively exhaustive. Attempting to ensure that all potential failures/risks are covered, the splits on each level should add up to 100% of the possibilities from the immediate node above. This could depend heavily on the way the splits are made. To help make this rigorous, each split must have a "basis" against which to judge this requirement. Typically split bases are physically or logically related to the way the system is organized, not opinions or brainstorm ideas. There are several ways of classifying failure groups by exploiting the objective nature of the system: Function flow, Lifecycle phases, Geometrical features, Time trend, Components, Material composition, etc. Such physical characteristics of a system are independent of the person doing the analysis.

The corresponding improvement in FTA can be demonstrated with the example mentioned in Principle #3 again. The proposed three stages for design cover 100% of the time regime a structure experiences during a full lifecycle. Following this principle, many other splits in Fig. 2 can also be reorganized.

Principle #5: Decompose each node until a physics level is reached where specific corrective actions can be implemented to raise reliability. In the end, only those nodes with the highest relative probability of occurrence require corrective or preventive actions.

In Fig. 2, "Environmental stress" is a very important branch which discovers the external energy threats for subsea pipeline reliability. But the bottom elements reached in this tree didn't reach the measureable level. To make improvements, for example, F27 "Excess vibration between pipe supports " can be split further into "direction", "frequency", "amplitude", "time profile" which could be physically measured or tested. This would also provide a parameters list for reliability designers.

Principle #6: Levels developed vertically should be as symmetrical as possible under each parallel node and should follow similar logical sequences and relationships. This provides level independence. If levels are interchanged in a tree, the final quantitative result will be insensitive to level number. This rule should be regarded as a guideline to organize the vertical development of a tree. Layers that are not interchangeable in a tree become sensitive to layer structure by virtue of amplifying the noise associated with probability assignments given to different layer arrangements. Then when the system fails it is commonly attributed to something "unexpected", or a "rare event" because it was improperly ignored or unforeseen. This principle serves only the quantitative calculation based on the new FET structure. Therefore this is not applicable in the FTA.

The nomenclature, "Failure Expansion Tree", FET, is chosen to be consistent with Principles#1 and #5 to emphasize the physical and geometrical nature of failures. The word "fault" suggests that human errors are also failure modes. A human's only role in a "fault" at the design stage is to overlook sources of potential physical failures. Human deficiencies can never be prevented, but physical failures may be prevented, managed or dealt with in some way. However, they must first be recognized. Of course, human error may result in system damage during manufacturing, installation or operation. However, such errors must be dealt with by using preventive methods such as Standard Operating Procedures (SOPs) [16] or Poka Yoke [17], not through fundamental physical design. A failure itself is <u>always</u> physical, no matter its originating chain-of-events. Strategies for interrupting a chain-of-events which leads to catastrophic failure, whether human originated or via some minor initiating event, will be the subject of a future paper.

These six principles assure a reliability design with the lowest probability of leakage. It forces people to think how to group things more consistently and logically. Thus, groupings in a FET are physically related and parallel to each other, as opposed to the less-controlled groupings which arise from brain storming and other subjective hypothesis generation methods. Lacking objectivity leads to ambiguous and inconsistent potential corrective actions.

RESULTS

Example of FET Application

Qualitative Analysis.

We give here an example of a portion of a tree built using the FET principles for one of the three major phases a subsea pipeline structure sees during its service lifecycle. In reliability design, there are three regimes which are generally considered to represent a full lifecycle of a structure: Infant mortality, Useful life and Wear out [18]. In spite of the suggestive names for these three regimes, it is not age that separates them; rather it is the relationship between energies in a system and the strength of a structure to resist those energies. Systems fail when energies exceed strength, whatever the units of energy and strength are and whenever these occur. Infant mortality suggests a weak structure coming from the manufacturing or installation process. Structures fail quickly after being put in the field because normal energies in the environment exceed initial structure strength. Useful life problems result from some kind of energy attack which exceeds the structure's design strength, resulting in malfunction or destruction. In Wear out, the strength of a structure gradually weakens as it gets older until it can no longer sustain normal environmental energies. It fails naturally towards the end of its life. In the example FET shown in Fig. 3, after making it clear that the focus of this analysis branch is under water, not above or at the water surface, the new tree begins by dividing failures into these three different life cycle regimes.

Failure Expansion Tree
Reliability Failure Mode FET for Subsea Pipeline System

		Objective: Improve FTA for subsea pipeline system			Last Update	3-Jan-14

Principles: Straight Observations; Think physics - Failures are physical - No prejudging; Splits: Mutually exclusive --> no overlap on the same level; Splits: Collectively exhaustive --> add up to 100% within the split basis

Level No.	Basis of Split	Improve the reliability for a subsea pipeline system			Problem Statement as management sees it: Division
1	Pipe location	Under the water `a`	Transmission air-to-water `a`	Above water `a`	Where
2	"ESD" Model	Infant mortality (initial strength) `b`	Useful life (Energy spike) `b`	Wear out (Strength decay) `b`	Which
3	Location of energy source	External energy sources (E.E.) `c`		Internal energy sources (I.E.) `c`	Where
4	Categories of energy sources	Natural environment (N.E.) `d1`		Human / Industrial activities (H.I.) `d1`	Which
5	Energy categories	Mechanical Spikes (N.E.) `e1`	Hydraulic Spikes (N.E.) `e1`	Thermal, Other (N.E.) `?`	Which
6	Surroundings	Floating ice impact `16`	Soil / seabed shift `15`	Other `?`	Which
5			Hydraulic Spikes (N.E.) `e1`		
6	External hydraulic pressure sources	Earthquake pressure wave `?`	Hurricane `17`	Other `?`	Which
4			Human / Industrial activities (H.I.) `d1`		
5	Energy categories	Mechanical Spikes (H.I) `e2`	Hydraulic Spikes (H.I.) `e2`	Chemical, Thermal, Other (H.I.) `?`	Which
6	H.I Mechanical impact sources	Boat/ fishing/ resident impact `18,20 21,23`	Marine construction activities `19, 22, 24`	Other `?`	Which
5			Hydraulic Spikes (H.I.) `e2`		
6	H.I hydraulic pressure sources	Boat/ fishing/ resident hydraulic pressure `?`	Explosion `f1`	Other `?`	Which
7	Explosive hydraulic sources	Military `?`	Terrorist `?`	Oil discovery, Other `?`	Which
3			Internal energy sources (I.E.) `c`		
4	Energy categories	Mechanical Spikes (I.E.) `d2`	Hydraulic Spikes (I.E.) `d2`	Chemical, Thermal, Other (I.E.) `d2`	Which
5	Internal mechanical impact sources	Large debris in the crude oil `e3`	Wedged internal scrubber `?`	Other `?`	Which
6	Physical features of debris	Hardness `?`	Shape `?`	Size `?`	Which
4			Hydraulic Spikes (I.E.) `d2`		
5	Internal hydraulic pressure sources	Oil valve / pump malfunction `?`		Other `?`	Which
4			Chemical, Thermal, Other (I.E.) `d2`		
5	I.E -thermal chemical sources	Crude oil temperature `?`	Chloride /sulfide PPM `?`	Other `?`	which

Figure 3: Failure Expansion Tree for identifying the risk factors in a subsea pipeline system during Useful life. Boxes crossed off with a dashed line would be considered in other branches of a complete FET. Failure codes on top of the boxes refer to nodes taken from Reference[3] (see Table S2) and are shown for comparison purposes only. Figures to the right of the boxes refer to probability data taken from [23] and used in the rank order analysis of Fig. 4. Question marks are failure modes unidentified in the original FTA analysis. We call these reliability design "leakage"; they are areas where rare events, being unforeseen, might occur.

Fig. 3 shows a partially expanded FET for the "Useful life" regime of a subsea pipeline. We select this regime for an example because it has the longest expected duration in time and is the regime where "rare events" are most likely to occur. An example of an FET for Infant Mortality is given elsewhere [19].

Construction of the tree and its calculation are simply done with a Microsoft Excel Spreadsheet. Down the left hand column of the tree we state the "Basis of Split", the expectation being that if the basis is clear, then it should be obvious whether one can enumerate all the possibilities defined by it. This is the application of Principle #4, Collectively Exhaustive. Inspection should also make it clear if Principle #3, Mutually Exclusive, has been applied. In the figure we can see how the original FTA nodes (listed above the boxes) have been divided and classified following the mutually exclusive and collectively exhaustive principles, and we can also see where items have been missed.

Examination of the tree reveals that sometimes it is not practical to explicitly list every possibility per Principle #4. For example, within "Energy categories", which is necessarily an open-ended basis of split, only those types of energies which are reasonably expected to be members of the box immediately above are listed. Other energy types are possible and we remind ourselves of this by including "Other" in the right-most box. However, such abbreviations should be used judiciously because it is important to remember that the primary objective of the FET principles is to force one to think about possibilities that might not otherwise come to mind.

As noted, because the branch expanded in Fig. 3 is Useful life, the focus is on excessive energies, that is, energy spikes which exceed design strength. In a dynamic marine environment, a normally functioning structure might see impacts from numerous such energy sources. In this regime, failure rate is associated with the frequency of occurrence of these sources. In the example we distinguish two geometrically separated energy origins, "External energy sources" and "Internal energy sources". "External energy sources" are then divided into "Natural environment (N.E.)" and "Human/industrial activities (H,I)". Under "Natural environment (N.E.)", the failure code F38 described as "Floating ice compressive strength" from the original FTA belongs to the "Floating ice impact" node in Fig. 3. Since the nature of ice flow is random in location and time (except for seasonality) the structure should ideally survive this kind of event/energy.

While it is clearly not cost-effective to make the pipeline infinitely strong to withstand all ice impacts, recognizing the existence and inability to control such a failure mode may lead one to include an "Inherent Safety" system [20] to limit the effect of any impact to a minor incident rather than allowing a rupture to continue unabated and causing a huge disaster. Such a system could also mitigate the effects of other rare or uncontrollable events, such as earthquakes, tsunamis, terrorism, etc., thus serving "double duty". Furthermore, such a system should be fail-safe, that is, not requiring a fully operational system in case more than one subsystem is compromised by an energy spike. Multi-subsystem failure was one of the reasons why the March 2011 earthquake in Japan turned the Fukushima Daiichi nuclear power plant failure into such a disaster [21].

We also see missed items in the original analysis. For comparison purposes, missed nodes are denoted with a "?" in Fig. 3. For example "Large debris carried by the crude oil" was not listed. Such "leakage" in the original FTA methodology is one reason why rare/unexpected events creep in. The new FET methodology with its related principles and careful decomposition seeks to solve this problem. Further decomposition of the Debris node provides an example of reaching the decomposition limit of Principle #5. While one cannot control the hardness or shape of unexpected debris in the oil, one could certainly limit its size with some kind of filter and diversion system to remove it, thus improving system reliability and breaking a potentially fatal chain-of-events. With this split under Useful Life we think of debris large enough to cause significant energy spikes from impacts inside the pipe. Following symmetry Principle #6, this same split would appear again under Wear Out where sand or other small particles would cause abrasive wear inside the pipe.

Note that in Fig. 3 all boxes relate to physical things in accordance with Principle #1. Absent are all references to management team quality, regulations, operator skill, poor inspection etc. Such items may be further up the chain of events that allow a failure to eventually develop, however the failure itself is always physical. We must first think of the physical nature of failures. Then either preventive action can be taken during design, or an up-stream system put in place to prevent the failure via some chain-of-events which may or may not involve human error. Again, the FET methodology focuses on the former, which must be recognized first, before the latter can be made effective.

Another significant difference between the FTA and a FET is that the FTA tries to identify not only bottom line elements, but also their interactions. The FET only attempts to find the bottom line individual physical elements. The reason for this is that interactions are often not possible to predict, and it is precisely such interactions that can lead to unexpected results. The philosophy with FET is that unknown interactions among the physical elements must be discovered experimentally through reliability testing using methods such as "multiple environment overstress testing" (MEOST) [22]. Other differences between FTA and FET are summarized in Table S1.

Quantitative Analysis.

As mentioned, the Useful Life regime was selected here for a calculation example because this is where rare events are most likely to occur. Clearly, there is never enough time or money during the design and installation phases to address all possible failure modes. The purpose of quantitative analysis is, therefore, to rank order the contributions of basic elements to the top event in terms of their relative likelihood of occurrence, and thus provide focus during design for most important reliability issues. Occurrence frequency data can be obtained from prior knowledge and experience in the field or, often helpfully, from other fields which may provide additional insight. Because of the way a failure tree is constructed when following FET Principle #3, there is only one kind of logic relationship involved, the "Or" gate. As shown in Fig. 4, relative probability calculations will therefore contribute multiplicatively through all intermediate levels up to the top level.

Rank Order Calculations

Branch Series	Factor Names (Level in FET →)	Source Reference Codes	Level 5 Raw	Level 5 Allocated	Allocated %	Contribution to the Top Event	Failure Rank
1	Under this water	a			33%		
	1-2 Useful life(Energy spike)	b			33%		
	1-2-1 External energy sources (E.E)	c			50%		
	1-2-1-1 Natural environment (N.E.)	d1			50%		
	1-2-1-1-1 Mechanical Spikes (N.E.)	e1			60%		
	1-2-1-1-1-1 Floating ice impact	16	2.3E-04	19%		2.83E-03	5
	1-2-1-1-1-2 Soil / seabed shift	15	9.9E-04	81%		1.13E-02	2
	1-2-1-1-1-3 Other	?	0	0%		0	6
	1-2-1-1-2 Hydraulic Spikes(N.E.)	e1			60%		
	1-2-1-1-2-1 Earthquake pressure wave	?	0	0%		0	6
	1-2-1-1-2-2 Hurricane	17	1.7E-03	100%		1.38E-02	1
	1-2-1-1-2-3 Other	?	0	0%		0	6
	1-2-1-1-3 Thermal, Other (N.E.)	?		0%		0	6
	1-2-1-2 Human / Industrial activities (H.I)	d1			50%		
	1-2-1-2-1 Mechanical Spikes (H.I)	e2			50%		
	1-2-1-2-1-1 Boat fishing/ resident impact	18	2.7E-03	11%		9.92E-03	3
		20	2.2E-03				
		21	4.1E-03				
		23	1.5E-03				
	1-2-1-2-1-2 Hydraulic Spikes (H.I)	19	1.3E-03	29%		4.07E-03	4
		22	8.5E-04				
		24	2.2E-03				
	1-2-1-2-1-3 Chemical, Thermal, Other (H.I)	?	0	0%		0	6
	1-2-1-2-2 Hydraulic Spikes(H.I)	e2			50%		
	1-2-1-2-2-1 Boat fishing/ resident hydraulic pressure	?	0	0%		0	6
	1-2-1-2-2-2 Explosion	f1	1	100%			
	1-2-1-2-2-2-1 Military	7	0	0%		0	6
	1-2-1-2-2-2-2 Terrorist	?	0	0%		0	6
	1-2-1-2-2-3 Oil discovery, Other	7	0	0%		0	6
	1-2-1-2-3 Other	?	0	0%		0	6
	1-2-1-2-3 Chemical, Thermal, Other (H.I)	?	0	0%		0	6
	1-2-2 Internal energy sources (I.E.)	c			50%		
	1-2-2-1 Mechanical Spikes (I.E.)	d2			33%		
	1-2-2-1-1 Large debris in the crude oil	e3			60%		
	1-2-2-1-1-1 Hardness	?	0	0%		0	6
	1-2-2-1-1-2 Shape	?	0	0%		0	6
	1-2-2-1-1-3 Size	?	0	0%		0	6
	1-2-2-1-2 Wedged internal scrubber	?		0%		0	6
	1-2-2-1-3 Other	?		0%		0	6
	1-2-2-2 Hydraulic Spikes(I.E.)	d2			33%		
	1-2-2-2-1 Oil valve / pump malfunction	?		0%		0	6
	1-2-2-2-2 Other	?		0%		0	6
	1-2-2-3 Chemical, Thermal, Other (I.E.)	d2			33%		
	1-2-2-3-1 Crude oil temperature	?		0%		0	6
	1-2-2-3-2 Chloride /sulfide PPM	?		0%		0	6
	1-2-2-3-3 Other	?		0%		0	6

Figure 4: Rank Order Analysis of Useful Life factors. Calculations for ranking are based on the organization of the FET in Fig. 3. Under Source Reference Codes, values are from [23], Table S3. Letters and "?" were not identified or evaluated by the original FTA, so corresponding probability data are unavailable. For demonstration purposes only, missing items are arbitrarily divided equally based on the number of nodes within a given level and "?" are denoted as "0". Obviously, proper values should be inserted by field experts or suitably researched.

Values for the nodes in Fig. 4 are taken either from the FTA example in [23] (Table S3) or they are arbitrarily divided equally within their level for demonstration purposes. Because the methodology requires nodes within a given level to be mutually exclusive and collectively exhaustive, the numerical values are always allocations of 100%, not absolute probabilities. Relative allocation is greatly facilitated by virtue of the fact that each node on a given level has the same "basis", so the elements are directly comparable to each other. After multiplicatively calculating top-level contributions, the values are rank ordered on the far right. Fig. 4 is an image cut directly from an Excel spreadsheet.

In practice, a Failure Expansion Tree (like a FTA) will expand very quickly as the decomposition progresses. During decomposition, probabilities would be updated for all the current bottom nodes and then further decomposition would focus only on those which are among the top, say, ~25% of all current bottom nodes. Rank order of the top dozen nodes or so would be kept. When those dozen are all down to controllable physical factors, then the decomposition is complete and corrective actions can be taken top-down on the rank ordered list using whatever time and resources are available.

A Normal (i.e., Gaussian) distribution is often assumed in failure probability distribution studies. However, a problem arises with this assumption. It grossly underestimates the probability of rare events whose tails follow a power law distribution [24]. The long power law tail results from the accumulation of all ignored, missed or improperly evaluated factors. The improved FET is more likely to list all the factors because of the way it is constructed and the ranking is more realistic because of the way probabilities are allocated. Thus, the close-in parts of the power law tail are more likely to be included in the top dozen ranking, and therefore dealt with during design. This results in improved reliability in the face of what would otherwise be considered a rare event.

Notice that quantitative analysis of the FET only covers the probability of occurrence of the top level event. In a large and complex system, multiple subprojects or subsystems should be identified and treated individually to make the analysis manageable. The objective of each subproject is to reduce occurrence of its own top level event.

Finally, for proper risk assessment, the ultimate decision of what should be improved in the design should be made based on the product of the calculated "Contribution to the top event" and its associated potential economic loss. Economic loss evaluation requires another round of analysis, not shown in Fig. 4, nor discussed in this paper.

Some additional comments about Fig. 4 are worth noting:

- According to the rank order calculation, the top 5 factors requiring the most attention are: "Hurricane", "Soil/seabed shift","Boat/fishing resident impact", "Hydraulic Spikes(H,I)", and "Floating ice impact". All of these areas can (and probably should) be divided further into measurable and controllable levels. For example, "Boat/fishing resident impact" could be divided in ways which might suggest specific methods to measure, detect and prevent occurrence. If failure is considered unavoidable, automatic rupture detection and shutdown systems should be incorporated to mitigate a potential disaster.

- The original FTA analysis in [23] concluded the top 5 factors were "Third party damage", "Corrosion", "Vortex-induced vibration", "Management", "Operation". While corrosion would belong to the "Wear Out" branch of a complete FET, the ranking is very different between the FET and the FTA. This is partly due to the different way the FET decomposition proceeds with its focus on the physics of failures, but in this example it is also largely due to the limited availability of data from prior work for the FET calculation. Where a "?" was encountered, the item is missing in the original FTA's and no attempt was made to quantify it. We simply denoted it as "0". Clearly, allocations must be developed for each FET level and node based on the frequency of occurrence of comparable nodes or other relative estimations.

- The Useful life branch of a complete FET is relatively short compared to branches of the other two regimes, mostly because there are fewer spike energy sources than there are ways things can go wrong during manufacturing, installation or Wear out. For example other energy sources such as biological, chemical/corrosive or abrasive would all fall under Wear out because they accumulate over time, but would not produce energy spikes. Initial strength weaknesses such as porous welds, dents, or low metallurgical hardness would fall under Infant mortality. Strength is not considered at all under Useful life because in this regime we only look for energies that exceed the design strength, not the actual strength. If normal energies exceed actual strength of some parameter which was not considered in the design, this would fall under Infant Mortality. Improved reliability would result from identifying this parameter, hopefully first with an Infant Mortality FET, and then strengthening it. However, multi-layer protective analysis would also call for use of techniques such as MEOST to identify unforeseen weaknesses should they be missed up front, even with a FET analysis.

Sensitivity of Quantitative Calculations.

Based on Principle #6, layers in the FET should be reasonably interchangeable without significantly affecting the quantitative calculation results, and therefore ranking. In other words, results should be relatively insensitive to tree layout as long as that layout follows a physical and logical sequence. The example in Fig. 5 demonstrates how this is achieved with a FET.

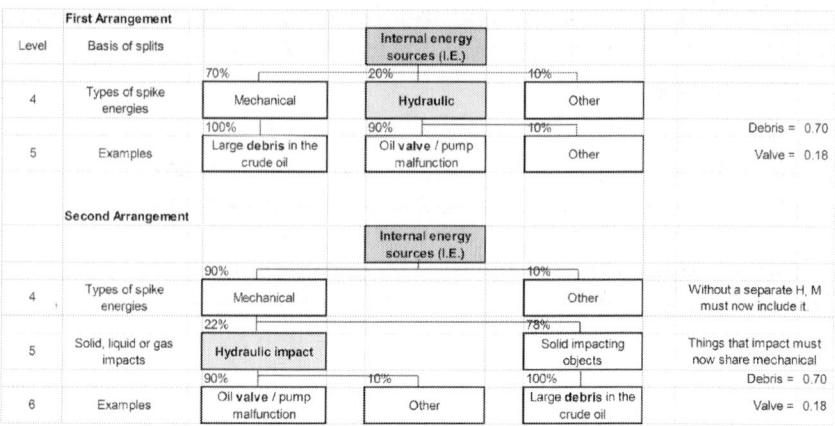

Figure 5: Demonstration of calculated results from two different tree structures with interchanged layers.On top of each box, the number is the proposed percent allocation with respect to the box above it. Node probability is the product of all layers above it and is calculated on the right hand side for Debris impact and Valve failures.

In the first arrangement "Hydraulic" is listed on level 4, parallel to "Mechanical" and "Other". It is rated as 20% of the total with respect to "Internal energy sources". In terms of frequency of occurrence, one might think of this data coming from experts having seen 7 mechanical events, 2 hydraulic events and 1 other kind of event, giving rise to the 70-20-10 percent split. Similarly, "experience" would have given rise to the other values shown. Under this tree arrangement, the "Debris" ranking score will be 100% × 70% = 0.7, while the "Valve" ranking score will be 90% × 20% = 0.18.

In the second arrangement, "Hydraulic" is considered a part of "Mechanical". It is now thought of as only one type of mechanical force (a conforming one) and would be listed in parallel with, say, "Solid impacting objects". This arrangement adds an intermediate layer between levels 4 and 6 with the split basis, "Solid, liquid or gas impacts" (the relevant possible states of impacting matter). "Mechanical" at level 4 must now be 90% since it subsumed "Hydraulic", while "Other" remains at 10% (or 9 events vs. 1 event in terms of frequency). "Hydraulic" itself was originally 2 events, while large debris was 7 events (100% of Mechanical in the first arrangement). These two add up to a total of 9 events. Normalizing the values to 100% gives us an allocation of 22% for Hydraulic and 78% for Solid Objects. The final calculation in this arrangement is "Debris" = 90% × 78% × 100% = 0.70 and "Valve" = 90% × 22% × 90% = 0.18, exactly the same as in the first arrangement.

There are two reasons that interchanging levels does not significantly affect the final ranking of each node. First, all the nodes in a given level must to add up to 100% through allocation. Second, where node data is available as a frequency of occurrence at the top level, that node's relative ranking with respect to other comparable nodes at the same level is always fixed, regardless of the level at which it appears in the FET.

Notably, the allocation approach within levels makes it easier to judge relative contributions for boxes for which factual data may not be readily available. Comparison of equivalent things is always easier than making absolute statements.

DISCUSSION

Table S1 compares nine different aspects of the two methodologies. The advantages and disadvantages of each methodology are shown in column 2 and 3. The 4th column lists which principles have been applied to make the improvements, and explains how they work.

CONCLUSIONS

A new technique for improving Fault Tree Analysis, FTA, has been presented and applied to improving submarine pipeline reliability during its useful life. The original FTA nodes from [3]were filtered, reconstructed, revised and extended by following a logical sequence of physical decomposition using six key principles aimed at reducing failure mode leakage, instead of brainstorming or use of other subjective risk factor identification methods. The improved tree is called a Failure Expansion Tree, or "FET", suggesting a focus on physics and geometry. The calculation of each risk factor's relative contribution to the top event is carried out with a Microsoft Excel Spreadsheet. No "least cut set" is needed for massive calculation as in FTA. Simple "Or" logic is used throughout the new tree, which largely reduces the complication of computer programming and clarifies failure routes. Besides helping one focus during the design phase, the structure also enhances the decision maker's ability to quickly review and react later during problem solving.

The six principles for building a Failure Expansion Tree are: (1) failures are things that are physically observable on parts, (2) avoid confounding cause and effect which often results from skipping over an intervening chain of events, (3) split boxes must be mutually exclusive to avoid overlap, (4) they must be collectively exhaustive and decomposed according to the physics and geometry of the structure to assure completeness, (5) nodes must be decomposed until corrective action can be taken at a physical level and (6) trees must have an element of symmetry to avoid quantitative sensitivity to level structure. A key advantage of the FET approach is to achieve more complete risk factor coverage, and thus uncover potential rare events. Rank order calculations allow focusing improvement work on the "Top 5" or so factors which have the highest relative probability of occurrence, or the highest cost if failure occurs. In the example presented, the FET approach was more likely to identify events caused by energy spikes during useful life. Addressing these will achieve improved reliability.

The example in this paper was for methodology illustration purposes only. A practical application with more comprehensive splits down to addressable physics levels and proper quantitative data would require suitable industry experts.

ACKNOWLEDGMENTS

The authors are grateful for the support from Prof. Jing Zhou of the Faculty of Infrastructure Engineering, Dalian University of Technology.

AUTHOR CONTRIBUTIONS

Conceived and designed the experiments: JL. Performed the experiments: JL. Analyzed the data: JL. Contributed reagents/materials/analysis tools: JL YY MZ. Wrote the paper: JL.

REFERENCES

1. Xu P, Zhu J (2007) Methodology and Application of Risk Analysis Theory. NUDT Publish House: 135–137. (in Chinese).

2. Rausand M, Hoylan A (2004) System Reliability Theory Models, Statistical Methods, and Applications. Wiley Series in probability and statistics - second edition. 50–59.

3. Xie Y, Yao A, Qian H (2006) The establishment and qualitative analysis of the fault tree for failure of submarine pipeline system. China Offshore Oil and Gas 18–3: 214–215 (in Chinese).

4. Morris DV (1991) Proceedings of the international Workshop of Offshore Pipeline Safety, A&M University, College Station, TX.

5. Peng D (2008) Research on Safety Reliability and Risk Assessment of Submarine Pipeline, China University of Petroleum. 116–119 (in Chinese).

6. Xie Y (2007) Research of Risk Assessment Technology for Subsea Oil and Gas Pipeline System. Southwest Petroleum University, 59–61 (in Chinese).

7. Muhlbauer WK (2004) Pipeline Risk Management Manual, Third edition.

8. Clifton Ericson, (1999) Fault Tree Analysis - A History. 17th International Systems Safety Conference.

9. Javadi MS, Nobakht A, Meskarbashee A (2011) Fault Tree Analysis Approach in Reliability Assessment of Power System. International Journal of Multidisciplinary Sciences and Engineering 2: 6.

10. Sarbes A (1990) Severe Accident Risks: An Assessment for Five U.S. Nuclear Power Plants. Washington, DC: U.S. Nuclear Regulatory Commission, NUREG–1150.

11. Process Safety Management of Highly Hazardous Chemicals, Explosives and Blasting Agents, Occupational Safety and Health Administration, Final Rule - February 24, 1992.

12. Dong Y, Yu D (2005) Estimation of failure probability of oil and gas transmission pipelines by fuzzy fault tree analysis. Journal of loss prevention in the process industries 18: 83–88 (in Chinese).. doi: 10.1016/j.jlp.2004.12.003

13. Wang Q, Zhao J (2008) (Mechanical and Power Engineering College of Nanjing University of Technology), NATUR GASIND. Vol. 28, 5 (in Chinese).

14. Peng k (2007) The Application of Human Error Analysis to Risk Assessment for Benthal Pipeline under Haphazard Load. Tianjin University: 36–40 (in Chinese).

15. BS 5760–5 (1991) Reliability of systems, equipment and components; Part 5: Guide to failure modes, effects and criticality analysis (FMEA and FMECA), British Standards Institution, London.

16. Hoyle D (2012) ISO 9000 Quality Systems Handbook - updated for the ISO 9001:2008 standard (sixth edition).

17. Shing S (1986) Zero quality control: Source inspection and the poka-yoke system. Productivity Press. 45p.

18. Rausand M, Hoylan A (2004) System Reliability Theory Models, Statistical Methods, and Applications. Wiley Series in probability and statistics - second edition. 10p.

19. Lin J (2013) New Development of FTA and Application to Reliability Failure Analysis, 2013 International Conference on Industrial Engineering and Management Science (ICIEMS 2013), Shanghai, China, September 28–29. in press.

20. Heikkila AM (1999) Inherent safety in process plant design: an index-based approach. VTT Technical Research Centre of Finland. 32–35.

21. Zhou Q (2012) Failure cause and effect analysis of Fukushima nuclear power plant, Popular Utilization of Electricity (1) (in Chinese).

22. Bhote KR, Bhote AK (2004) World class reliability: Using multiple environment overstress tests to make it happen. AMACOM Div American Management Association. 155–189.

23. Xie Y (2007) Research of Risk Assessment Technology for Subsea Oil and Gas Pipeline System, Southwest Petroleum University: 12–13, 23–24 (in Chinese).

24.   Gabaix X, Gopikrishnan P, Plerou V, Stanley HE (2006) Institutional investors and stock market volatility. The Quarterly Journal of Economics 121(2): 3–5. doi: 10.1162/qjec.2006.121.2.461

Citations

CHAPTER 1

Mikael Gustavsson, Mohammad Shahriari, and Mats Lindgren, "Evaluating EML Modeling Tools for Insurance Purposes: A Case Study," International Journal of Chemical Engineering, vol. 2010, Article ID 104370, 13 pages, 2010. doi:10.1155/2010/104370.

CHAPTER 2

K. K. Salam, A. O. Alade, A. O. Arinkoola, and A. Opawale, "Improving the Demulsification Process of Heavy Crude Oil Emulsion through Blending with Diluent," Journal of Petroleum Engineering, vol. 2013, Article ID 793101, 6 pages, 2013. doi:10.1155/2013/793101

CHAPTER 3

Qingmin Hou, Liang Ren, Wenling Jiao, Pinghua Zou, and Gangbing Song, "An Improved Negative Pressure Wave Method for Natural Gas Pipeline Leak Location Using FBG Based Strain Sensor and Wavelet Transform," Mathematical Problems in Engineering, vol. 2013, Article ID 278794, 8 pages, 2013. doi:10.1155/2013/278794.

CHAPTER 4

Chinedu I. Ossai, "Advances in Asset Management Techniques: An Overview of Corrosion Mechanisms and Mitigation Strategies for Oil and Gas Pipelines," ISRN Corrosion, vol. 2012, Article ID 570143, 10 pages, 2012. doi:10.5402/2012/570143

CHAPTER 5

Nascimento, A. , Coelho-Gomes, C. , Barbarino, E. and Lourenço, S. (2014) Temporal Variations of the Chemical Composition of Three Seaweeds in Two Tropical Coastal Environments. Open Journal of Marine Science, 4, 118-139. doi: 10.4236/ojms.2014.42013.

CHAPTER 6

Rizzi, V. , Longo, A. , Fini, P. , Semeraro, P. , Cosma, P. , Franco, E. , García, R. , Ferrándiz, M. , Núñez, E. , Gabaldón, J. , Fortea, I. , Pérez, E. and Ferrándiz, M. (2014) Applicative Study (Part I): The Excellent Conditions to Remove in Batch Direct Textile Dyes (Direct Red, Direct Blue and Direct Yellow) from Aqueous Solutions by Adsorption Processes on Low-Cost Chitosan Films under Different Conditions. Advances in Chemical Engineering and Science, 4, 454-469. doi: 10.4236/aces.2014.44048.

CHAPTER 7

Susan Caines, Faisal Khan, John Shirokoff, Analysis of pitting corrosion on steel under insulation in marine environments, Journal of Loss Prevention in the Process Industries, Volume 26, Issue 6, November 2013, Pages 1466-1483, ISSN 0950-4230, http://dx.doi.org/10.1016/j.jlp.2013.09.010.

CHAPTER 8

Lin J, Yuan Y, Zhang M (2014) Improved FTA Methodology and Application to Subsea Pipeline Reliability Design. PLoS ONE 9(3): e93042. doi:10.1371/journal.pone.0093042.

Index